83 *General irreducible Markov chains and non-negative operators*

ESA NUMMELIN
Associate Professor of Applied Mathematics,
University of Helsinki

General irreducible Markov chains and non-negative operators

CAMBRIDGE UNIVERSITY PRESS
Cambridge
London New York New Rochelle
Melbourne Sydney

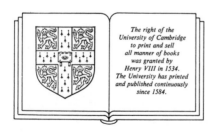

Published by the Press Syndicate of the University of Cambridge
The Pitt Building, Trumpington Street, Cambridge CB2 1RP
32 East 57th Street, New York, NY 10022, USA
296 Beaconsfield Parade, Middle Park, Melbourne 3206, Australia

First published 1984

Printed in Great Britain at the University Press, Cambridge

Library of Congress catalogue card number: 83-23995

British Library cataloguing in publication data

Nummelin, Esa
General irreducible Markov chains and
non-negative operators.—(Cambridge
tracts in mathematics; 83)
1. Markov processes
I. Title
519.2′33 QA274.7

ISBN 0 521 25005 6

TM

To Leena, Mikko and Anna

Contents

Preface

The basic object of study in this book is the theory of discrete-time Markov processes or, briefly, Markov chains, defined on a general measurable space and having stationary transition probabilities.

The theory of Markov chains with values in a countable set (discrete Markov chains) can nowadays be regarded as part of classical probability theory. Its mathematical elegance, often involving the use of simple probabilistic arguments, and its practical applicability have made discrete Markov chains standard material in textbooks on probability theory and stochastic processes.

It is clear that the analysis of Markov chains on a general state space requires more elaborate techniques than in the discrete case. Despite these difficulties, by the beginning of the 1970s the general theory had developed to a mature state where all of the fundamental problems – such as cyclicity, the recurrence–transience classification, the existence of invariant measures, the convergence of the transition probabilities – had been answered in a satisfactory manner. At that time also several monographs on general Markov chains were published (e.g. Foguel, 1969*a*; Orey, 1971; Rosenblatt, 1971; Revuz, 1975).

The primary motivation for writing this book has been in the recent developments in the theory of general (irreducible) Markov chains. In particular, owing to the discovery of embedded renewal processes, the 'elementary' techniques and. constructions based on the notion of regeneration, and common in the study of discrete chains, can now be applied in the general case.

Our second motivation is to point out the close connections between the theories of Markov chains and non-negative operators (operators induced by a non-negative transition kernel). This relationship is analogous to that between discrete Markov chains and non-negative matrices (cf. the monograph by Seneta (1981)). Since the emphasis here lies on Markov chains we shall discuss the theory of general non-negative transition kernels only as far as it naturally arises as an extension of the theory of markovian transition kernels. However, even within this relatively narrow scope, we are able to develop some fundamental concepts and results, among them a general Perron–Frobenius-type theory for non-negative kernels. (For an account of the general, functional analytic approach to non-negative operators the reader is referred to the monograph by Schaefer (1974).)

Chapter 1 contains basic definitions and some preliminary results on kernels and Markov chains. In Chapter 2 we examine the fundamental concepts of irreducibility and cyclicity.

Chapter 3 deals with the concepts of transience and recurrence and the associated decomposition results. First we shall analyse general irreducible kernels and prove Vere-Jones' and Tweedie's theorem which classifies them as R-recurrent or R-transient kernels. The remainder of Chapter 3 deals with Markov chains. Starting from elementary potential theoretic notions we end up with Hopf's decomposition theorem, stating that the state space of a Markov chain can be split into a 'transient' and a 'recurrent' part.

Chapter 4 is concerned with renewal processes and renewal sequences embedded in general irreducible Markov chains and non-negative kernels. The techniques introduced in Chapter 4 form a basic tool in the proofs and constructions of the later chapters.

Chapter 5 deals with the principal 'eigenfunctions' and 'eigenmeasures' (to be called here R-invariant functions and R-invariant measures, respectively) of a non-negative kernel, and with the related concepts of R-positive and R-null recurrence. A systematic study is also made of the degrees of (positive) recurrence for Markov chains; in Chapter 6 these are then used to give results regarding the rates of convergence of the transition probabilities.

Finally, Chapters 6 and 7 develop the limit theory of the iterates of non-negative kernels and Markov chain transition probabilities. Chapter 6 is devoted to total variation convergence results, among them Orey's fundamental convergence theorem and various results concerning the rate of convergence. Chapter 7 contains miscellaneous limit theorems, including results on the convergence of sums and ratios of transition probabilities and central limit theorems.

Our approach is based on the use of probabilistic notions and arguments. Most of the concepts and results, even if first formulated for general, non-markovian kernels, will be interpreted in terms of Markov chain sample paths. It is assumed that the reader is familiar with basic concepts of probability theory, such as random variables, conditional expectations, etc. The reader interested only in operator theory can read Sections 1.1, 2.1–2.4, 3.1–3.2, 4.1–4.3, 5.1–5.2 and 6.2, skipping everywhere those parts where Markov chains (and probability theory in general) are discussed. (However, for 6.2 the renewal theorems of the preceding Section 6.1 should be consulted. Although Sections 5.7 and 6.7 deal with general non-markovian kernels, the preceding 'probabilistic' sections form a necessary background for them.)

The book contains a few recurring examples (indicated by the letters (*a*)–(*k*)), the primary aim of which is to illustrate the general theoretical concepts and results in certain special cases. Some examples refer to practical applications of general Markov chains. In most cases the detailed calculations leading to the statements and results in the examples are left to the reader as exercises.

The references which have been used can be found in the bibliography. They are discussed in a separate section with title 'Notes and comments'. We do not intend to give a systematic account of the historical development of the concepts and results presented in this book. In order to trace this development more exactly, the reader is advised to study the references and their bibliographies.

There are many people to whom I am indebted for their support and criticism. My special thanks are due to Elja Arjas for many comments and suggestions which helped to improve the quality of the text. I also wish to thank Richard Tweedie for important comments on an early version. For helpful remarks I am grateful to Juha Ahtola, Heikki Bonsdorff, Priscilla Greenwood, Seppo Niemi and Karen Simon. For the typing of the manuscript I thank Raili Pauninsalo. The support received from the Academy of Finland and from the Emil Aaltonen Foundation is also gratefully acknowledged.

Helsinki, June 1983 Esa Nummelin

1

Preliminaries

In this chapter we introduce the basic terminology and definitions concerning Markov chains and non-negative kernels.

1.1 Kernels

Let E be a set and $\mathscr{E}$ a σ-algebra of subsets of E. We assume that the σ-algebra $\mathscr{E}$ is *countably generated*, i.e. generated by a countable collection of subsets of E. The measurable space $(E, \mathscr{E})$ is called the *state space* and the points of E are called *states*. The symbol $\mathscr{E}$ will also be used to denote the collection of extended real valued measurable functions on $(E, \mathscr{E})$. The symbols $x, y, \ldots$ denote states, $A, B, \ldots$ denote elements of the σ-algebra $\mathscr{E}$, and $f, g, \ldots$ denote extended real valued measurable functions on $(E, \mathscr{E})$.

We write $\mathscr{M}$ for the collection of signed measures on $(E, \mathscr{E})$. The symbols $\lambda, \mu, \ldots$ denote elements of $\mathscr{M}$. When $\mathscr{A}$ is any collection of functions or signed measures, we write $\mathscr{A}_+$ (resp. b$\mathscr{A}$) for the class of non-negative (resp. bounded) elements of $\mathscr{A}$. In what follows, we shall refer to the elements of $\mathscr{M}_+$ simply as measures. $\mathscr{M}^+$ denotes the class of positive measures, i.e. $\mathscr{M}^+ = \{\lambda \in \mathscr{M}_+ : \lambda(E) > 0\}$. The elements of b$\mathscr{M}$ will also be called finite signed measures.

Definition 1.1. A (non-negative) *kernel* on $(E, \mathscr{E})$ is any map $K: E \times \mathscr{E} \to \bar{\mathbb{R}}_+$ satisfying the following two conditions:

(i) for any fixed set $A \in \mathscr{E}$, the function $K(\cdot, A)$ is measurable;

(ii) for any fixed state $x \in E$, the set function $K(x, \cdot)$ is a measure on $(E, \mathscr{E})$.

A kernel K is called *σ-finite*, if there exists an $\mathscr{E} \otimes \mathscr{E}$-measurable function $f \in \mathscr{E} \otimes \mathscr{E}$, $f > 0$ everywhere, such that

$$\int K(x, \mathrm{d}y) f(x, y) < \infty \quad \text{for all } x \in E;$$

finite, if

$$K(x, E) < \infty \quad \text{for all } x \in E;$$

bounded, if

$$\sup_{x \in E} K(x, E) < \infty;$$

substochastic, if

$$K(x, E) \leq 1 \quad \text{for all } x \in E;$$

and *stochastic*, if

$$K(x,E)=1 \quad \text{for all } x\in E.$$

Examples 1.1. (*a*) Suppose that E is *discrete*, that means a countable set. Let $\mathscr{E}$ be the σ-algebra of all subsets of E. Then any kernel K on $(E,\mathscr{E})$ can be identified with the non-negative matrix

$$k(x,y)\stackrel{\text{def}}{=}K(x,\{y\}), \quad x,y\in E.$$

K is (for example) σ-finite if and only if the matrix elements $k(x,y)$ are finite.

(*b*) Let φ be a σ-finite measure on $(E,\mathscr{E})$ and let k be a non-negative $\mathscr{E}\otimes\mathscr{E}$-measurable function. The kernel

$$K(x,\mathrm{d}y)\stackrel{\text{def}}{=}k(x,y)\varphi(\mathrm{d}y)$$

is called an *integral kernel* (with basis φ and density k). Clearly, K is σ-finite if and only if, for all $x\in E$, $k(x,y)$ is finite for φ-almost all $y\in E$.

Any kernel K can be interpreted as a non-negative linear operator on the cone $\mathscr{E}_+$ by defining

$$Kf(x)=\int K(x,\mathrm{d}y)f(y), \quad f\in\mathscr{E}_+.$$

(When integration is over the whole state space E, we usually omit the symbol E indicating the integration area.) Similarly, K acts as an operator on $\mathscr{M}_+$:

$$\lambda K(A)=\int\lambda(\mathrm{d}x)K(x,A), \quad \lambda\in\mathscr{M}_+.$$

We denote by I the identity kernel $I(x,A), x\in E, A\in\mathscr{E}$:

$$I(x,A)\stackrel{\text{def}}{=}1_A(x)\stackrel{\text{def}}{=}\begin{cases}1 & \text{if } x\in A,\\ 0 & \text{if } x\notin A.\end{cases}$$

If K_1 and K_2 are two kernels, their *product kernel* K_1K_2 is defined by

$$K_1K_2(x,A)=\int K_1(x,\mathrm{d}y)K_2(y,A).$$

The *iterates* $K^n, n\geq 0$, of a kernel K are defined by setting $K^0=I$, and iteratively,

$$K^n=KK^{n-1} \quad \text{for } n\geq 1.$$

Henceforth we make the following:

Basic assumption. K is a fixed non-negative kernel on $(E,\mathscr{E})$. All the iterates $K^n, n\geq 1$, of K are σ-finite.

In the case where K is substochastic, we shall use the symbol P instead of K, and call $K = P$ a *transition probability*. Note that in this case, trivially, all the iterates $P^n, n \geq 1$, are substochastic (whence σ-finite).

1.2 Markov chains

Let $(\Omega, \mathscr{F})$ be a measurable space, to be called the *sample space*, and let $\mathbb{P}$ be a probability measure on $(\Omega, \mathscr{F})$. A measurable map ξ defined on $(\Omega, \mathscr{F})$ and taking values in an arbitrary measurable space $(\Omega', \mathscr{F}')$ is called an Ω'-valued *random element*. An $\bar{\mathbb{R}}$-valued random element is called a *random variable*. A sequence $(\xi_n; n \geq 0)$ of Ω'-valued random elements is called an Ω'-valued *stochastic process*. Note that the stochastic process $(\xi_n; n \geq 0)$ can also be viewed as an $(\Omega')^{\times \infty}$-valued random element.

If ξ is an Ω'-valued random element, we shall write $\mathscr{L}(\xi)$ for the probability distribution of ξ, i.e.,

$$\mathscr{L}(\xi)(A') = \mathbb{P}\{\xi \in A'\}, \quad A' \in \mathscr{F}'.$$

The analogous notation will be used for conditional distribution.

Let $\mathscr{F}_0 \subseteq \mathscr{F}_1 \subseteq \cdots \subseteq \mathscr{F}_n \subseteq \cdots$ be an increasing sequence of sub-σ-algebras of $\mathscr{F}$, to be called a *history*. A stochastic process (ξ_n) is said to be *adapted to the history* $(\mathscr{F}_n)$, if ξ_n is $\mathscr{F}_n$-measurable for each $n \geq 0$. Note that a stochastic process (ξ_n) is always adapted to the history

$$\mathscr{F}_n^\xi = \sigma(\xi_0, \ldots, \xi_n), \quad n \geq 0,$$

called the *internal history of* (ξ_n).

A Markov chain $(X_n; n \geq 0)$ is an E-valued stochastic process having the *Markov property*; that means, at every n, the next state X_{n+1} depends only on the present state X_n of the process. *We suppose for a while that $K = P$ is a stochastic transition probability.*

Definition 1.2. An E-valued stochastic process $(X_n; n \geq 0)$ is called a *Markov chain with transition probability* P, provided that

$$\mathscr{L}(X_{n+1} | \mathscr{F}_n^X) = \mathscr{L}(X_{n+1} | X_n) = P(X_n, \cdot) \quad \text{for all } n \geq 0. \tag{1.1}$$

(We shall omit systematically the phrase 'almost surely' when dealing with equalities involving conditional distributions.)

A Markov chain (X_n) is called *Markov with respect to a history* $(\mathscr{F}_n)$, provided that it is adapted to $(\mathscr{F}_n)$, and (1.1) holds when $\mathscr{F}_n^X$ is replaced by $\mathscr{F}_n$, i.e.,

$$\mathscr{L}(X_{n+1} | \mathscr{F}_n) = \mathscr{L}(X_{n+1} | X_n) = P(X_n, \cdot) \quad \text{for all } n \geq 0. \tag{1.2}$$

Sometimes, if we want to emphasize that (X_n) is Markov w.r.t. a general history $(\mathscr{F}_n)$ we will speak of the *Markov chain* $(X_n, \mathscr{F}_n)$.

The distribution $\mathscr{L}(X_0)$ of X_0 is the *initial distribution* of the Markov chain (X_n). If $\mathscr{L}(X_0)=\varepsilon_x$, i.e., $\mathbb{P}\{X_0=x\}=1$, for some state x, then x is called the *initial state* of the chain.

Examples 1.2. (*a*) Let $(X_n;n\geq 0)$ be a Markov chain on a discrete state space E (abbreviated 'a *discrete Markov chain*') with transition probability P; P can be identified with the matrix

$$\begin{aligned} p(x,y) &\stackrel{\text{def}}{=} P(x,\{y\}) \\ &= \mathbb{P}\{X_{n+1}=y \mid X_n=x\}, \quad x,y\in E, n\geq 0, \end{aligned}$$

called the *transition matrix* of (X_n) (cf. Example 1.1(*a*)). We have

$$p^m(x,y)\stackrel{\text{def}}{=} P^m(x,\{y\})=\mathbb{P}\{X_{n+m}=y \mid X_n=x\}, \quad x,y\in E, m,n\geq 0.$$

(*c*) Let $z_n, n=1,2,\ldots$, be i.i.d. (abbreviation for 'independent, identically distributed'), finite random variables with common distribution F,

$$F(A)=\mathbb{P}\{z_n\in A\}, \quad A\in\mathscr{R}.$$

Let Z_0 be a finite random variable, independent of $\{z_n;n\geq 1\}$, with distribution F_0. Define

$$Z_n=Z_0+\sum_1^n z_m, \quad \text{for } n\geq 1.$$

The stochastic process $(Z_n;n\geq 0)$ is called a *random walk* (on $\mathbb{R}$). Clearly, it is a Markov chain on $(\mathbb{R},\mathscr{R})$ with initial distribution $\mathscr{L}(Z_0)=F_0$ and transition probability

$$P(x,A)=F(A-x), \quad x\in\mathbb{R}, A\in\mathscr{R}$$

$$(A-x\stackrel{\text{def}}{=}\{y-x: y\in A\}).$$

(*d*) Let $z_n, n\geq 1$, and Z_0 be as in the above example. Suppose that $Z_0\geq 0$ a.s. (abbreviation for 'almost surely'). Set $W_0=Z_0$, and

$$W_n=(W_{n-1}+z_n)_+ \quad \text{for } n\geq 1.$$

The stochastic process $(W_n;n\geq 0)$ is called a *reflected random walk* (on $\mathbb{R}_+$). It is a Markov chain on $(\mathbb{R}_+,\mathscr{R}_+)$ with initial distribution $\mathscr{L}(W_0)=F_0$ and transition probability

$$P(x,A)=1_A(0)F((-\infty,-x))+F(A-x), \quad x\in\mathbb{R}_+, \; A\in\mathscr{R}_+.$$

(In *queueing theory* W_n is commonly interpreted as the waiting time of the nth customer (before service). This follows if we set $z_n=s_{n-1}-(T_n-T_{n-1})$, where s_n is the service time of the nth customer and T_n is the arrival epoch of the nth customer, and assume that $s_0,s_1,\ldots$ are i.i.d., and T_1-T_0, $T_2-T_1,\ldots$ are i.i.d., independent of $\{s_n;n\geq 0\}$.)

(*e*) Let $z_n, n \geq 1$, and Z_0 be as in Example (*c*). Suppose that they all are non-negative (a.s.). Then the random walk (Z_n) is called a *renewal process on* $\mathbb{R}_+$. For any $t \geq 0$, write

$$V_t^+ = \inf\{Z_n - t\colon Z_n \geq t, n \geq 0\}.$$

The continuous-time stochastic process $(V_t^+; t \geq 0)$ is called the *forward process* associated with the renewal process $(Z_n; n \geq 0)$. For any $\delta > 0$, the *skeletons* $(V_{n\delta}^+; n \geq 0)$ form a Markov chain on $(\mathbb{R}_+, \mathscr{R}_+)$ with initial distribution F_0.

(*f*) Let $z_n, n \geq 1$, and Z_0 be as in Example (*c*). Let ρ be a constant. Set $R_0 = Z_0$, and iteratively

$$R_n = \rho R_{n-1} + z_n, \quad n \geq 1.$$

The stochastic process $(R_n; n \geq 0)$, called an *autoregressive process*, is a Markov chain on $(\mathbb{R}, \mathscr{R})$.

(*g*) Examples (*c*), (*d*) and (*f*) are special cases of the following scheme, which could be called a *stochastic difference equation*: Let $(E', \mathscr{E}')$ be an arbitrary measurable space, and let $f: E \times E' \to E$ be jointly measurable. Let $z_n, n \geq 1$, be i.i.d., E'-valued random elements, and let X_0 be an E-valued random element, independent of $(z_n; n \geq 1)$. Setting iteratively

$$X_n = f(X_{n-1}, z_n), \quad n \geq 1,$$

we obtain a Markov chain on $(E, \mathscr{E})$.

In the above stochastic difference equation the Markov property is retained if we allow z_n to depend on the previous state X_{n-1}:

$$\mathscr{L}(z_n | \mathscr{F}_{n-1}^X \vee \mathscr{F}_{n-1}^z) = \mathscr{L}(z_n | X_{n-1}) = P'(X_{n-1}, \cdot)$$

for some transition probability P' from $(E, \mathscr{E})$ into $(E', \mathscr{E}')$. The resulting Markov chain (X_n) represents an abstract *learning model* if we interpret X_n as the *states of learning* and the random variables z_n as *events* (induced by the states of learning through the transition probability P'). (See e.g. Norman, 1972).

In what follows, we shall refer to Examples (*a*)–(*g*) by using the above lettering.

Let us again return to the general theory. If (X_n) is a Markov chain with initial distribution $\mathscr{L}(X_0) = \lambda$ and transition probability P, then clearly

$$\mathbb{P}\{X_0 \in A_0, \ldots, X_n \in A_n\} = \int_{A_0} \lambda(\mathrm{d}x_0) \int_{A_1} P(x_0, \mathrm{d}x_1) \ldots \int_{A_n} P(x_{n-1}, \mathrm{d}x_n) \quad (1.3)$$

for all $n \geq 0, A_0, \ldots, A_n \in \mathscr{E}$. In particular, we have

$$\mathscr{L}(X_n) = \lambda P^n.$$

Conversely, if an E-valued stochastic process (X_n) satisfies (1.3) for all $n \geq 0, A_0, \ldots, A_n \in \mathscr{E}$, then it is a Markov chain with initial distribution λ and transition probability P.

For any given probability measure λ and stochastic transition probability P on $(E, \mathscr{E})$ we can always construct a Markov chain with initial distribution λ and transition probability P: Set $\Omega = E^{\times \infty}, \mathscr{F} = \mathscr{E}^{\otimes \infty}$, and define $X_n(\omega) = \omega_{n'}$, the $(n+1)$th coordinate of $\omega = (\omega_0, \omega_1, \ldots) \in \Omega$. On the cylinder sets $A_0 \times \cdots \times A_n$ of $\mathscr{F}$ define a probability measure $\mathbb{P}$ according to the formulas (1.3) and extend it to $\mathscr{F}$ to obtain the desired Markov chain. The Markov chain, constructed in this manner, is called the *canonical Markov chain* corresponding to the initial distribution λ and transition probability P. The sample space $(\Omega, \mathscr{F}) = (E^{\times \infty}, \mathscr{E}^{\otimes \infty})$ is the *canonical sample space.* For details of this construction see e.g. Doob (1953), Suppl., §2, Neveu (1965), Sect. V.1, or Revuz (1975), Ch.1, §2.

If the transition probability P is *not stochastic* then we proceed as follows: Take a point Δ not belonging to the set E and adjoin it to E to obtain the *extended state space* $(E_\Delta, \mathscr{E}_\Delta) = (E \cup \{\Delta\}, \sigma(\mathscr{E}, \{\Delta\}))$. Extend λ to $(E_\Delta, \mathscr{E}_\Delta)$ by setting $\lambda(\{\Delta\}) = 0$. The transition probability P is extended to $(E_\Delta, \mathscr{E}_\Delta)$ by setting

$$P(x, \{\Delta\}) = 1 - P(x, E), \quad x \in E,$$
$$P(\Delta, \{\Delta\}) = 1.$$

Clearly, this extended P is a stochastic kernel on $(E_\Delta, \mathscr{E}_\Delta)$ so that we can construct the corresponding canonical Markov chain on $(E_\Delta, \mathscr{E}_\Delta)$. The point Δ is called the *cemetery* of the Markov chain (X_n).

In what follows, whenever $K = P$ is a transition probability we shall think of it as the transition probability of a Markov chain (X_n). When P is not stochastic, we automatically make the extension described above.

In the sequel we will often use the notation $\mathbb{P}_\lambda$ (resp. $\mathbb{P}_x$) instead of $\mathbb{P}$ to indicate a specific initial distribution λ (resp. initial state x) of the chain. More generally, if $\lambda \in \mathscr{M}_+$ is an arbitrary measure on $(E, \mathscr{E})$, we write $\mathbb{P}_\lambda$ for the measure $\int \lambda(dx) \mathbb{P}_x(\cdot)$ on $(\Omega, \mathscr{F})$. The symbol $\mathbb{E}$ denotes the expectation operator corresponding to the probability measure $\mathbb{P}$.

If (X_n) is Markov w.r.t. a history $(\mathscr{F}_n)$, the formula (1.2) expresses in its simplest form the characteristic *Markov property* of the Markov chain $(X_n, \mathscr{F}_n)$. Below we shall formulate the Markov property in a slightly more general form stating that, given X_n, the whole post-n-chain $(X_n, X_{n+1}, \ldots)$ is conditionally independent of the past $\mathscr{F}_n$. For this we need the concepts of a shift operator and a functional.

A measurable map θ on the sample space $(\Omega, \mathscr{F})$ is called a *shift operator* (for the Markov chain (X_n)) provided that

$$X_n(\theta\omega) = X_{n+1}(\omega) \quad \text{for all } \omega \in \Omega, n \geq 0.$$

The iterates $\theta_m, m \geq 0$, of θ are defined by setting $\theta_0 = I_\Omega$, the identity operator on Ω, and iteratively,

$$\theta_m = \theta \circ \theta_{m-1} \quad \text{for } m \geq 1.$$

A random variable ζ, which is measurable w.r.t the σ-algebra $\mathcal{F}^X \stackrel{\text{def}}{=} \sigma(X_n; n \geq 0)$, i.e. of the form

$$\zeta = \eta(X_0, X_1, \ldots), \quad \eta \quad \mathcal{E}^{\otimes \infty}\text{-measurable},$$

is called a *functional* (of the Markov chain (X_n)).

Note that, for any $n \geq 0$, if ζ is a functional then the functional $\zeta \circ \theta_n$ is measurable w.r.t. the σ-algebra $\sigma(X_n, X_{n+1}, \ldots)$, and conversely, if a functional ζ' is measurable w.r.t. $\sigma(X_n, X_{n+1}, \ldots)$ then $\zeta' = \zeta \circ \theta_n$ for some functional ζ.

Theorem 1.1. (The Markov property). Let $(X_n; n \geq 0)$ be a Markov chain w.r.t. a history $(\mathcal{F}_n)$. Then for any non-negative functional ζ,

$$\mathbb{E}[\zeta \circ \theta_n | \mathcal{F}_n] = \mathbb{E}_{X_n}[\zeta] \quad \text{for all } n \geq 0.$$

Proof. When ζ is the indicator of a cylinder,

$$\zeta = 1_{\{X_0 \in A_0, \ldots, X_m \in A_m\}},$$

the result is a straightforward consequence of (1.2) and (1.3). The extension to the general case follows the standard lines. □

Note that Theorem 1.1 implies the existence of a regular version of the conditional distribution $\mathcal{L}(X_n, X_{n+1}, \ldots | \mathcal{F}_n)$, since

$$\mathcal{L}(X_n, X_{n+1}, \ldots | \mathcal{F}_n; X_n = x) = \mathcal{L}(X_0, X_1, \ldots | X_0 = x)$$
$$= \mathbb{P}_x \quad (\text{restricted to } \mathcal{F}^X).$$

2

Irreducible kernels

The main theme of this chapter is, broadly speaking, to investigate the 'communication structure' induced by the kernel K on the state space $(E, \mathscr{E})$. That is, we shall be concerned with the relation $x \to A$ on $E \times \mathscr{E}$, defined by

$$x \to A \quad \text{if and only if } K^n(x, A) > 0 \text{ for some } n \geq 1.$$

When $x \to A$, we say that the *set* $A \in \mathscr{E}$ *is attainable from the state* $x \in E$. If A is not attainable from x, i.e., $K^n(x, A) = 0$ for all $n \geq 1$, then we write $x \not\to A$. When $K = P$ is the transition probability of a Markov chain (X_n), we have the following probabilistic interpretation for attainability:

$$x \to A \quad \text{if and only if } \mathbb{P}_x\{X_n \in A \text{ for some } n \geq 1\} > 0.$$

2.1 Closed sets

First we shall define and study the so-called closed sets. These are the sets $F \in \mathscr{E}$ such that the complement F^c is not attainable from F.

Definition 2.1. A non-empty set $F \in \mathscr{E}$ is called *closed* (for the kernel K), if

$$K(x, F^c) = 0 \quad \text{for all } x \in F.$$

A set $B \in \mathscr{E}$ is called *indecomposable* (for K), if there do not exist two disjoint closed sets $F_1, F_2 \subset B$. A non-empty set $F \in \mathscr{E}$ is called *absorbing* (for K), if

$$K(x, F) = K(x, E) = 1 \quad \text{for all } x \in F.$$

Clearly, $F \in \mathscr{E}$ is closed if and only if $x \not\to F^c$ for all $x \in F$.

Note also that an absorbing set is always closed. Conversely, F may be closed without being absorbing: it is possible that

$$K(x, F) = K(x, E) \neq 1 \quad \text{for some } x \in F.$$

When F is closed, we can regard K also as a kernel on the restricted state space $(F, \mathscr{E} \cap F)$,

$$K|_F(x, A) \overset{\text{def}}{=} K(x, A), \quad x \in F, \ A \in \mathscr{E} \cap F.$$

$K|_F$ is called the *restriction of* K *to the closed set* F. Clearly, the iterates of $K|_F$ coincide with those of K on F,

$$(K|_F)^n = (K^n)|_F.$$

Similarly we can *restrict* K *to the complement* F^c *of the closed set* F,

$$K|_{F^c}(x, A) \overset{\text{def}}{=} K(x, A), \quad x \in F^c,\ A \in \mathscr{E} \cap F^c.$$

Again we have

$$(K|_{F^c})^n = (K^n)|_{F^c}.$$

Note that the restriction of K to an absorbing set is a stochastic transition probability.

The kernel G given by

$$G = \sum_0^\infty K^n$$

is called the *potential kernel* (of K). It may happen that G is not σ-finite, since it is possible that $G(x, A), x \in E, A \in \mathscr{E}$, admits only the values 0 and ∞. In any case we have the following:

Proposition 2.1. For any $n \geq 1$:

$$G = \sum_0^{n-1} K^m + K^n G = \sum_0^{n-1} K^m + GK^n.$$

In particular,

$$G = I + KG = I + GK,$$

and for any $f \in \mathscr{E}_+$,

$$\lim_{n \to \infty} \downarrow K^n G f = 0 \quad \text{on } \{Gf < \infty\}.$$

(The notation $\{Gf < \infty\}$ means the set $\{x \in E : Gf(x) < \infty\}$.)

Proof. Obvious. □

Let us denote by A^0 the set of states in A^c from which A is not attainable:

$$A^0 = \{x \in A^c : x \not\rightarrow A\} = \{G1_A = 0\}.$$

Proposition 2.2. For any $A \in \mathscr{E}: A^0$ is either empty or closed.

Proof. For all $x \in A^0 : KG(x, A) \leq G(x, A) = 0$, and hence $K(x, (A^0)^c) = 0$. □

Before proceeding further, we have to consider the notion of equivalence for kernels:

Two measures $\lambda, \mu \in \mathscr{M}_+$ are said to be equivalent, and we write $\lambda \sim \mu$ if they are mutually absolutely continuous, i.e. have the same null sets.

A σ-finite kernel $\tilde{K}$ is said to be *equivalent* to the kernel K, if for all $x \in E$, the measures $K(x,\cdot)$ and $\tilde{K}(x,\cdot)$ are equivalent.

Lemma 2.1. There exists a substochastic kernel $\tilde{K}, \tilde{K} \leq K$, which is equivalent to K. Moreover, there exists an $\mathscr{E} \otimes \mathscr{E}$-measurable version $\tilde{f}, 0 < \tilde{f} \leq 1$ everywhere, of the Radon–Nikodym derivative

$$\tilde{f}(x,y) = \frac{\tilde{K}(x,\mathrm{d}y)}{K(x,\mathrm{d}y)}.$$

The iterated kernels K^n and $\tilde{K}^n$ are equivalent, for any $n \geq 0$.

Proof. Let $f \in \mathscr{E} \otimes \mathscr{E}, f > 0$ everywhere, be such that the function $Kf \in \mathscr{E}_+$ defined by

$$Kf(x) = \int K(x,\mathrm{d}y) f(x,y), \quad x \in E,$$

is finite. Without any loss of generality we may suppose that $f \leq 1$. It is easy to check that the desired kernel $\tilde{K}$ is given by $\tilde{K}(x,\mathrm{d}y) = \tilde{f}(x,y)K(x,\mathrm{d}y)$, where

$$\tilde{f}(x,y) = 1_{\{K(x,E)>0\}}(Kf(x))^{-1} f(x,y) + 1_{\{K(x,E)=0\}}.$$

That K^n and $\tilde{K}^n$ are equivalent is obvious. □

Let $\psi \in \mathscr{M}^+$ be a σ-finite measure such that ψK is absolutely continuous w.r.t. ψ. There always exist such measures: Take any σ-finite measure $\varphi \in \mathscr{M}^+$ and set

$$\psi = \sum_0^\infty 2^{-(n+1)} \tilde{\varphi} \tilde{K}^n, \tag{2.1}$$

where $\tilde{\varphi}$ is a probability measure equivalent to φ and $\tilde{K}$ is a substochastic kernel equivalent to K.

Lemma 2.2. (i) If $\psi K \ll \psi$ then $\psi G \sim \psi$.

(ii) If $\psi K \ll \psi$ and E is indecomposable, then $\psi(F) > 0$ for all closed sets $F \in \mathscr{E}$.

Proof. (i) By induction $\psi K^n \ll \psi$ for all $n \geq 0$, and hence $\psi G = \sum_0^\infty \psi K^n \ll \psi$. On the other hand, trivially, $\psi \leq \psi G$.

(ii) By Proposition 2.2 and by the hypothesis F^0 is empty whenever F is closed. Hence $\psi G(F) > 0$, which by (i) implies $\psi(F) > 0$. □

Suppose now that $K = P$ is the transition probability of a Markov chain (X_n). A set F is called closed for (X_n), if it is closed for the transition probability P of (X_n). Similarly, we shall speak about indecomposable and absorbing sets for (X_n), and about the restriction of (X_n) to a closed set F or to the complement F^c. The latter, i.e. the restriction of (X_n) to the complement F^c of a closed set F, means that the set $\Delta_F \overset{\text{def}}{=} F + \{\Delta\}$ is made into a cemetery for (X_n).

Note that F is closed for (X_n) if and only if

$$\mathbb{P}_x\{X_n \in F^c \quad \text{for some } n \geq 1\} = 0 \quad \text{for all } x \in F;$$

F is absorbing for (X_n) if and only if

$$\mathbb{P}_x\{X_n \in F \text{ for all } n \geq 1\} = 1 \quad \text{for all } x \in F.$$

2.2 φ -irreducibility

Let $\varphi \in \mathscr{M}^+$ be a σ-finite measure on $(E, \mathscr{E})$, and let $B \in \mathscr{E}$ be a *φ-positive set* (i.e., a set with $\varphi(B) > 0$).

Definition 2.2. The set B is called *φ-communicating* (for the kernel K) if every φ-positive subset $A \subseteq B$ is attainable from B, i.e.

$$x \to A \quad \text{for all } x \in B \text{ and all } \varphi\text{-positive } A \subseteq B.$$

If the whole state space E is *φ-communicating*, then the kernel K is called *φ-irreducible*. K is called *irreducible*, if it is φ-irreducible for some φ. In this case the measure φ is called an *irreducibility measure* for K.

Examples 2.1. (*a*) When E is discrete, irreducibility (in the sense of the above definition) means (using an obvious notation) that for some non-empty subset $B \subseteq E$,

$$x \to y \quad \text{for all } x \in E,\ y \in B.$$

The corresponding irreducibility measure is the counting measure on B, that is the measure Card_B defined by

$$\text{Card}_B(A) = \text{Card}(A \cap B), \quad A \subseteq E.$$

(In matrix theory, irreducibility usually means Card-irreducibility, i.e.

$$x \to y \quad \text{for all } x, y \in E.)$$

(*b*) Suppose that K is the integral kernel with basis φ and density k. Then for every $n \geq 1$, K^n is the integral kernel with basis φ and density $k^{(n)}$, where $k^{(1)}(x, y) = k(x, y)$, and

$$k^{(n)}(x, y) = \int k(x, z) k^{(n-1)}(z, y) \varphi(\mathrm{d}z) \quad \text{for } n \geq 2.$$

Now, if there exists a φ-positive set $B \in \mathscr{E}$ such that for all $x \in E$,

$$\sum_1^\infty k^{(n)}(x, y) > 0 \quad \text{for } \varphi\text{-almost all } y \in B,$$

then K is φI_B-irreducible.

(*c*) (The random walk). Let ℓ denote the Lebesgue measure on $(\mathbb{R}, \mathscr{R})$. A distribution F is called *spread-out* if some convolution power F^{*n_0} of F is not singular w.r.t. ℓ. From the well known theorem of analysis, according to which the convolution of a bounded and of an ℓ-integrable function on $(\mathbb{R}, \mathscr{R})$ is continuous, it follows that if F is spread-out then there is an interval $[a, b], a < b$, and a constant $\beta > 0$ such that

$$F^{*2n_0}(\mathrm{d}t) \geq \beta \mathrm{d}t \quad \text{on } [a, b].$$

(We write $\ell(\mathrm{d}t) = \mathrm{d}t$.) The random walk (Z_n) is ℓ-irreducible if and only if the increment distribution F is spread-out.

(*d*) (The reflected random walk). Let ε_0 denote the probability measure on $(\mathbb{R}_+, \mathscr{R}_+)$ assigning unit mass to the origin. The reflected random walk (W_n) is ε_0-irreducible if and only if

$$\mathbb{P}\{z_1 < 0\} > 0, \text{ i.e., } F(\mathbb{R}_+) < 1.$$

(*e*) (The forward process). Let $F(t) = F([0, t]), t \geq 0$, and $\bar{M} = \text{ess sup } z_1 = \sup\{t : F(t) < 1\}$. Let ℓ_a denote the restriction of the Lebesgue measure to $[0, a)$, $0 < a \leq \infty$. Suppose that F is spread-out. Then, for any $\delta > 0$, the Markov chain $(V^+_{n\delta}; n \geq 0)$ is $\ell_{\bar{M}}$-irreducible.

(*f*) (The autoregressive process). Suppose that $|\rho| < 1$. Also suppose that the distribution F of z_n is not singular (w.r.t. ℓ), i.e., there is $f \in \mathscr{R}_+$ with $\int_{\mathbb{R}} f(t)\mathrm{d}t > 0$ such that

$$F(\mathrm{d}t) \geq f(t)\mathrm{d}t.$$

Then there is an interval Γ of positive length such that the autoregressive process $(R_n; n \geq 0)$ is ℓI_Γ-irreducible. (Hint: Consider first the case where $0 \leq \rho < 1$ and there is an interval $[a, b], a < b$, such that $f > 0$ on $[a, b]$. In this case, prove that the interval Γ defined by $\Gamma = [(1-\rho)^{-1}a, (1-\rho)^{-1}b]$ is ℓ-communicating, and then, that $x \to \Gamma$ for all $x \in \mathbb{R}$. After this consider the general case; observe that the 2-step chain $(R_{2n}; n \geq 0)$,

$$R_{2n} = \rho^2 R_{2(n-1)} + \rho z_{2n-1} + z_{2n}, \quad n \geq 1,$$

satisfies the above hypotheses.)

Let us denote

$$B^+ = B \cup \{x \in E : x \to B\} = \{G1_B > 0\}.$$

Recall from Proposition 2.2 that the complement $(B^+)^c = B^0$ is either empty or closed. Hence we can always restrict the kernel K to B^+.

We need the following simple lemma:

Lemma 2.3. Let $A, B \in \mathscr{E}$ and $x \in B^+$ be arbitrary. If $y \to A$ for all $y \in B$ then $x \to A$.

Proof. Easy □

Proposition 2.3. (i) If B is φ-communicating then it is indecomposable.
(ii) If B is φ-communicating then $K|_{B^+}$, the restriction of K to B^+, is φI_B-irreducible.

Proof. (i) Suppose that there were two disjoint closed sets in B, say F_1 and F_2. Then either $B \backslash F_1$ or $B \backslash F_2$ would be φ-positive. But this leads to a contradiction with the hypothesis.
(ii) Take any φ-positive set $A \subseteq B$. Then apply Lemma 2.3. □

According to part (ii) of the above proposition, if there exists a φ-communicating set B in the state space – and we are not interested in that part of the state space from which B is not attainable – then there is no loss of generality in assuming that K is irreducible.

It is clear that any measure ψ which is absolutely continuous w.r.t. an irreducibility measure φ is itself an irreducibility measure. Conversely, as we will see, starting from an irreducibility measure φ, we can construct a *maximal irreducibility measure*, that is an irreducibility measure ψ such that all other irreducibility measures are absolutely continuous w.r.t. ψ. Note that, by definition, a maximal irreducibility measure is unique up to the equivalence of measures.

Proposition 2.4. Suppose that K is φ-irreducible. Then:
(i) There exists a maximal irreducibility measure.
(ii) An irreducibility measure ψ is maximal if and only if $\psi K \ll \psi$.
(iii) Let ψ be a maximal irreducibility measure. If $\psi(B)=0$ then also $\psi(B^+)=0$.

Proof. Suppose that ψ is an irreducibility measure satisfying $\psi K \ll \psi$. By Lemma 2.2(i), $\psi G \ll \psi$. From this and from the φ-irreducibility it follows that if $\varphi(A)>0$ then $\psi(A)>0$. Hence ψ is maximal.

Conversely, if ψ is an irreducibility measure then so is ψK, and hence, if ψ is maximal then $\psi K \ll \psi$. By Lemma 2.2(i) we have also (iii).

It remains to prove (i). Define a finite measure ψ by (2.1). Clearly ψ is an irreducibility measure satisfying $\psi K \ll \psi$ and it is maximal by (ii). □

Examples 2.2. (*a*) Suppose that K is a Card_B-irreducible matrix, $B \subseteq E$ (cf. Example 2.1(*a*)). Set

$$F = \{y \in E : x \to y \text{ for some } x \in B\}.$$

Note that by Card_B-irreducibility, $B \subseteq F$. Now Card_F, the counting measure on F, is a maximal irreducibility measure. F is closed; moreover, it is the minimal closed set.

In what follows, when we deal with an irreducible matrix, F stands for the unique closed set having the properties stated above.

(*b*) Suppose that K is a φI_B-irreducible integral kernel with basis φ and density k (cf. Example 2.1(*b*)). Let B' denote the set

$$B' = B \cup \{y \in E : \sum_1^\infty \int_B \varphi(\mathrm{d}x) k^{(n)}(x, y) > 0\}.$$

Then $\psi = \varphi I_{B'}$ is a maximal irreducibility measure for K.

If K is irreducible (with maximal irreducibility measure ψ), we call a set $B \in \mathscr{E}$ *full* (for the kernel K) whenever $\psi(B^c) = 0$. It turns out that full and closed sets almost coincide:

Proposition 2.5. Suppose that K is irreducible with maximal irreducibility measure ψ. Then:

(i) Every closed set F is full.
(ii) If B is a full set then there exists a closed set $F \subseteq B$.
(iii) If $F_i, i \geq 1$, are closed then so is their intersection $\bigcap_1^\infty F_i$.

Proof. (i) $\psi(F^c) > 0$ leads to a contradiction with the hypothesis.

(ii) Set $F = (B^c)^0 = \{G1_{B^c} = 0\}$. By Propositions 2.2 and 2.4(iii), F is closed.

(iii) In general, the intersection of a countable number of closed sets is either empty or closed. But now, under the assumption of irreducibility, every F_i is full and hence so is their intersection $\bigcap_1^\infty F_i$. □

The result of Proposition 2.5 is very useful. When we want to prove that a given set B is full, we need to prove that B is closed or, at least, that it contains a closed set F.

2.3 The small functions

In this section we shall define and study an important class of functions, the so-called small functions. Later we will see that, in some respects, they play a similar role for a general kernel K, as do individual states in the case where E is discrete (and K is a matrix).

For the rest of this chapter we suppose that K is an irreducible kernel. ψ denotes a fixed maximal irreducibility measure. For any subclass $\mathscr{A}_+ \subseteq \mathscr{E}_+$ of non-negative measurable functions on $(E, \mathscr{E})$, $\mathscr{A}^+$ denotes the subclass of ψ-positive elements in $\mathscr{A}_+$:

$$\mathscr{A}^+ = \{f \in \mathscr{A}_+ : \psi(f) > 0\},$$

$$\left(\psi(f) \stackrel{\text{def}}{=} \int f(x)\psi(\mathrm{d}x) \stackrel{\text{def}}{=} \int_E f(x)\psi(\mathrm{d}x)\right).$$

When $1_A \in \mathscr{A}^+$, i.e., $\psi(A) > 0$, we write simply $A \in \mathscr{A}^+$.

We say that the kernel K *satisfies the minorization condition* $M(m_0, \beta, s, \nu)$, where $m_0 \geq 1$ is an integer, $\beta > 0$ a constant, $s \in \mathscr{E}^+$ a function, and $\nu \in \mathscr{M}^+$ a measure, if

$$K^{m_0}(x, A) \geq \beta s(x)\nu(A) \quad \text{for all } x \in E,\ A \in \mathscr{E},$$

or, briefly, writing $s \otimes \nu$ for the kernel $s \otimes \nu(x, A) = s(x)\nu(A)$, $x \in E, A \in \mathscr{E}$,

$$K^{m_0} \geq \beta s \otimes \nu. \tag{2.2}$$

Definition 2.3. A *function* $s \in \mathscr{E}^+$ is called *small* (for the kernel K), if K satisfies $M(m_0, \beta, s, \nu)$ for some $m_0 \geq 1, \beta > 0$, $\nu \in \mathscr{M}^+$. A *measure* $\nu \in \mathscr{M}^+$ is called *small*, if K satisfies $M(m_0, \beta, s, \nu)$ for some $m_0 \geq 1, \beta > 0, s \in \mathscr{E}^+$.

A *set* $C \in \mathscr{E}^+$ is called *small*, if its indicator 1_C is a small function, i.e., there are $m_0 \geq 1$, $\beta > 0$ and $\nu \in \mathscr{M}^+$ such that

$$K^{m_0}(x, \cdot) \geq \beta \nu(\cdot) \quad \text{for all } x \in C.$$

We shall use the symbol $\mathscr{S}^+$ to denote the class of small functions. In what follows, the symbols s and ν will stand exclusively for a small function and a small measure, respectively.

Examples 2.3. (*a*) Let K be a Card_F-irreducible matrix. Then any singleton $\{x_0\}, x_0 \in F$, is a small set. (K satisfies the minorization condition $M(1, 1, 1_{\{x_0\}}, K(x_0, \cdot))$.)

(*b*) Suppose that K is a φI_B-irreducible integral kernel with basis φ and density k. If there exist $m_0 \geq 1, \beta > 0$, and φ-positive sets $C, D \in \mathscr{E}, C \subseteq B$, such that

$$k^{(m_0)}(x, y) \geq \beta \quad \text{for all } x \in C,\ y \in D,$$

then C is a small set. (The corresponding minorization condition is $M(m_0, \beta, 1_C, \varphi I_D)$.)

(*d*) (The reflected random walk). When $F(\mathbb{R}_+) < 1$, the singleton $\{0\}$ is a small set.

(*e*) (The forward process). Suppose that F is spread-out and that $\delta > 0$ is such that $F[0, \delta) < 1$. Then there are an integer $m_0 \geq 1$ and a constant $\beta > 0$ such that $M(m_0, \beta, 1_{[0,\delta)}, \ell_\delta)$ holds, i.e., the interval $[0, \delta)$ is a small set.

(*f*) With the assumptions of Example 2.1(*f*) every bounded set $C \in \mathscr{R}$ such that $\ell(C \cap \Gamma) > 0$ is small. (Hint: Take $\varepsilon > 0$ to be a small constant, and choose N so big that $|\rho|^N |x| < \varepsilon$ whenever $x \in C$. Then observe that

$$R_N = z_N + \rho z_{N-1} + \cdots + \rho^{N-1} z_1 + \rho^N R_0.$$

By using the hypothesis that F is non-singular we can find an interval $[c, d], c < d$, and a constant $\beta > 0$ such that

$$P^N(x, \mathrm{d}y) \geq \beta \mathrm{d}y \quad \text{whenever } x \in C,\ y \in [c, d].)$$

Remarks 2.1. (i) A small measure ν is always an irreducibility measure for K. Hence it is absolutely continuous w.r.t. ψ.

(ii) By 'multiplying' both sides of (2.2) by K^m we see that $K^m s$ is small, for all $m \geq 0$. Similarly, νK^m is small for all $m \geq 0$. Hence, by irreducibility, in most cases there is no loss of generality in assuming that $\nu(s) > 0$.

(iii) A small function or measure remains small when multiplied by a constant $\gamma > 0$. Consequently, in most cases one can simply assume that $\beta = 1$.

(iv) For any constant $\gamma > 0$, the set $C = \{s \geq \gamma\}$ is small whenever it is ψ-positive.

(v) If $K = P$ is a transition probability, then ν is finite, and s is bounded by $(\beta\nu(E))^{-1}$. Then, without loss of generality one can assume that

$$P^{m_0} \geq s \otimes \nu, \text{ where } 0 \leq s \leq 1, \text{ and } \nu \text{ is a probability measure.} \tag{2.3}$$

It is by no means clear that any small functions or measures should exist. However:

Theorem 2.1. Suppose that K is an irreducible kernel. Then $\mathscr{S}^+ \neq \emptyset$.

In order to prove this theorem we need some new definitions and preliminary results.

Recall that the measurable space $(E, \mathscr{E})$ was assumed to be countably generated. Then there is a sequence $(\mathscr{E}_i; i \geq 1)$ of finite partitions of E which generate $\mathscr{E}$. We can assume that $\mathscr{E}_{i+1}$ is finer than $\mathscr{E}_i$, i.e. for every i, any member of $\mathscr{E}_i$ is a finite union of sets from $\mathscr{E}_{i+1}$. Hence $\sigma(\mathscr{E}_i) \subseteq \sigma(\mathscr{E}_{i+1})$ for every i. For $x \in E$, let E_x^i be the unique member of $\mathscr{E}_i$ which includes the state x. Let λ and φ be two finite measures on $(E, \mathscr{E})$, and let

$$\lambda(\mathrm{d}y) = \lambda_{\mathrm{a}}(\mathrm{d}y) + \lambda_{\mathrm{s}}(\mathrm{d}y)$$

be the usual Lebesgue decomposition of λ into the absolutely continuous part λ_{a} and the singular part λ_{s} (w.r.t. φ).

The first lemma is the basic *differentiation theorem of measures*. For a proof, see e.g. Doob (1953), Ch. 7, §8.

Lemma 2.4.

$$\lim_{i \to \infty} 1_{\{\varphi(E_x^i) > 0\}} \frac{\lambda(E_x^i)}{\varphi(E_x^i)} = \frac{\mathrm{d}\lambda_a}{\mathrm{d}\varphi}(x) \quad \text{for } \varphi\text{-a.e. } x \in E. \qquad \square$$

Being a σ-finite measure for each $x \in E$, $K(x,\cdot)$ admits a Lebesgue decomposition

$$K(x, \mathrm{d}y) = k(x, y)\varphi(\mathrm{d}y) + K_{\mathrm{s}}(x, \mathrm{d}y),$$

where for any fixed $x \in E$, $k(x,\cdot) \in \mathscr{E}_+$ and $K_{\mathrm{s}}(x,\cdot)$ is singular w.r.t. φ. It is not immediately clear that we can choose the density $k = k(x, y)$ to be jointly measurable.

Lemma 2.5. There exists a non-negative $\mathscr{E} \otimes \mathscr{E}$-measurable version of the density k.

Proof of Lemma 2.5. By Lemma 2.1 it is no restriction to assume that the kernel K is substochastic. Now the functions $k_i \in \mathscr{E} \otimes \mathscr{E}$ defined by

$$k_i(x, y) = 1_{\{\varphi(E_y^i) > 0\}} \frac{K(x, E_y^i)}{\varphi(E_y^i)}$$

are non-negative and $\mathscr{E} \otimes \mathscr{E}$-measurable, and by Lemma 2.4 their limit $\lim_{i \to \infty} k_i$ is a version of the density k. □

We note in passing the following corollary of Lemma 2.5:

Example 2.4. (*b*) K is an integral kernel with basis φ if (and only if)

$$K(x, \cdot) \ll \varphi \quad \text{for all } x \in E.$$

Lemma 2.6. There exist non-negative $\mathscr{E} \otimes \mathscr{E}$-measurable versions of the densities $k^{(n)}$ of the iterates K^n, which satisfy

$$\begin{aligned} k^{(m+n)}(x, z) &\geq \int K^m(x, \mathrm{d}y) k^{(n)}(y, z) \\ &\geq \int k^{(m)}(x, y) k^{(n)}(y, z) \varphi(\mathrm{d}y) \end{aligned}$$

for all $m, n \geq 1, x, z \in E$.

Proof of Lemma 2.6. Let $k_0^{(n)}$ be a non-negative $\mathscr{E} \otimes \mathscr{E}$-measurable version of the density of K^n. Define $k^{(n)} \in \mathscr{E} \otimes \mathscr{E}$, $n \geq 1$, by setting $k^{(1)} = k_0^{(1)}$ and, iteratively,

$$\begin{aligned} k^{(n)}(x, z) &= k_0^{(n)}(x, z) \\ &\vee \max_{1 \leq m \leq n-1} \int K^{(m)}(x, \mathrm{d}y) k^{(n-m)}(y, z) \quad \text{for } n \geq 2. \end{aligned}$$

It is easy to check that $k^{(n)}, n \geq 1$, satisfy the given requirements. □

The following lemma is the key to the proof of Theorem 2.1. Let $A, B \in \mathscr{E} \otimes \mathscr{E}$ be arbitrary, and let $A_1(x)$ and $B_2(z)$ denote the sections

$$A_1(x) = \{y \in E : (x, y) \in A\},$$
$$B_2(z) = \{y \in E : (y, z) \in B\}.$$

The composition $A\circ B\in\mathscr{E}\otimes\mathscr{E}\otimes\mathscr{E}$ of A and B is defined by

$$\begin{aligned}A\circ B&=(A\times E)\cap(E\times B)\\&=\{(x,y,z)\in E\times E\times E:(x,y)\in A,\ (y,z)\in B\}.\end{aligned}$$

Write φ^n for the product measure $\varphi\times\cdots\times\varphi$ (n times).

Lemma 2.7. If $\varphi^3(A\circ B)>0$, then there exist φ-positive sets $C,D\in\mathscr{E}$ such that

$$\gamma\overset{\text{def}}{=}\inf_{x\in C,\,z\in D}\varphi(A_1(x)\cap B_2(z))>0.$$

Proof of Lemma 2.7. Set $E^i_{x,y}=E^i_x\times E^i_y$. By Lemma 2.4 there are φ^2-null sets $N_1,N_2\in\mathscr{E}\otimes\mathscr{E}$ such that

$$\lim_{i\to\infty}\frac{\varphi^2(A\cap E^i_{x,y})}{\varphi^2(E^i_{x,y})}=1\quad\text{for all }(x,y)\in A\backslash N_1$$

and

$$\lim_{i\to\infty}\frac{\varphi^2(B\cap E^i_{y,z})}{\varphi^2(E^i_{y,z})}=1\quad\text{for all }(y,z)\in B\backslash N_2.$$

Fix a triplet $(u,v,w)\in(A\backslash N_1)\circ(B\backslash N_2)$ and an integer j big enough so that

$$\frac{\varphi^2(A\cap E^j_{u,v})}{\varphi^2(E^j_{u,v})}\geq\frac{3}{4}\ \text{and}\ \frac{\varphi^2(B\cap E^j_{v,w})}{\varphi^2(E^j_{v,w})}\geq\frac{3}{4}.\tag{2.4}$$

Let

$$\begin{aligned}C&=\{x\in E^j_u:\varphi(A_1(x)\cap E^j_v)\geq\tfrac{3}{4}\varphi(E^j_v)\},\\D&=\{z\in E^j_w:\varphi(B_2(z)\cap E^j_v)\geq\tfrac{3}{4}\varphi(E^j_v)\}.\end{aligned}$$

It follows easily from (2.4) that $\varphi(C)>0$ and $\varphi(D)>0$. For any $x\in C$ and $z\in D$ we have

$$\varphi(A_1(x)\cap B_2(z))\geq\tfrac{1}{2}\varphi(E^j_v)>0\quad\text{by (2.4)}.\qquad\square$$

Now we are able to prove Theorem 2.1:

Proof of Theorem 2.1. Let φ be a probability measure which is equivalent to ψ. By irreducibility, we have for any $x\in E$

$$\sum_1^\infty k^{(m)}(x,y)>0\quad\text{for }\varphi\text{-a.e. }y\in E.$$

It follows that there exist integers $m_1,m_2\geq 1$ such that

$$\iiint k^{(m_1)}(x,y)k^{(m_2)}(y,z)\varphi(\mathrm{d}x)\varphi(\mathrm{d}y)\varphi(\mathrm{d}z)>0,$$

which in turn implies that for $\delta>0$ sufficiently small, the composition $A\circ B$

of the sets

$$A = \{k^{(m_1)} \geq \delta\}, \quad B = \{k^{(m_2)} \geq \delta\},$$

is φ^3-positive. If C, D and γ are as in Lemma 2.7 then by Lemma 2.6, for all $x \in C$, $z \in D$:

$$k^{(m_1+m_2)}(x, z) \geq \int_{A_1(x) \cap B_2(z)} k^{(m_1)}(x, y) k^{(m_2)}(y, z) \varphi(\mathrm{d}y) \geq \gamma\delta^2.$$

It follows that K satisfies the minorization condition $M(m_0, \beta, s, v)$ with $m_0 = m_1 + m_2, \beta = \gamma\delta^2, s = 1_C$ and $v = \varphi I_D$, and thus C is a small set. □

The following two results will be needed in the sequel. The first is a sharpening of Theorem 2.1:

Proposition 2.6. For any set $B \in \mathscr{E}^+$, there exists a small set $C \in \mathscr{S}^+$ with $C \subseteq B$.

Proof. By Remark 2.1(iv) it suffices to prove the existence of a small function $s \in \mathscr{S}^+$ with $\{s > 0\} \subseteq B$. To this end, let C' be a small set. By irreducibility, there is $m \geq 1$ such that the function $s = I_B K^m 1_{C'}$ belongs to $\mathscr{E}^+$. By Remark 2.1(ii) s is small. □

Proposition 2.7. (i) For any $f \in \mathscr{E}^+$ and any small function $s \in \mathscr{S}^+$, there exist an integer $m \geq 1$ and a constant $\gamma > 0$ such that

$$K^m f \geq \gamma s. \tag{2.5}$$

In particular, for any $f \in \mathscr{E}^+$ and any small set $C \in \mathscr{S}^+$, there exists an integer $m \geq 1$ such that

$$\inf_C K^m f > 0.$$

(ii) For any $f \in \mathscr{E}^+$ and $s \in \mathscr{S}^+$, there exists a constant $\gamma > 0$ such that

$$Gf \geq \gamma Gs.$$

(iii) For any $x \in E$ and any small measure $v \in \mathscr{M}^+$, there exists a constant $\gamma > 0$ such that

$$G(x, \cdot) \geq \gamma v G.$$

Proof. (i) We have

$$K^m f \geq \beta(v K^{m-m_0} f) s \quad \text{for all } m \geq m_0.$$

Choose $m \geq m_0$ such that $\gamma = \beta v K^{m-m_0} f > 0$.

(ii) 'Multiply' both sides of (2.5) by K^n and sum n over $\mathbb{N}$.

(iii) The proof is similar to that of (ii). □

2.4 Cyclicity

Next we shall examine the cyclic behaviour of an irreducible kernel. *We assume that K is an irreducible kernel satisfying the minorization condition* $M(m_0, \beta, s, \nu)$.

Definition 2.4. A sequence $(E_0, E_1, \ldots, E_{m-1})$ of m non-empty disjoint sets in $\mathscr{E}$ is called an *m-cycle* (for the kernel K), provided that for all $i = 0, \ldots, m-1$, and all $x \in E_i$:

$$K(x, E_j^c) = 0 \quad \text{for } j = i + 1 \pmod{m}.$$

Note that, if the sets $E_0, \ldots, E_{m-1}$ form an m-cycle, then their union $E_0 + \cdots + E_{m-1}$ is closed (whence by Proposition 2.5(i) also full). Also note that for all $n \geq 1$, $i = 0, \ldots, m-1$, $x \in E_i$:

$$K^n(x, E_j^c) = 0 \quad \text{for } j = i + n \pmod{m};$$

in particular,

$$K^m(x, E_i^c) = 0,$$

i.e. E_i is a closed set for the kernel K^m.

It turns out that the small function s cannot be strictly positive on two different sets E_i and E_j of a cycle.

Proposition 2.8. Let $(E_0, \ldots, E_{m-1})$ be an m-cycle and let N be the ψ-null set $N = (E_0 + \cdots + E_{m-1})^c$. Then there is an index i, $0 \leq i < m$, such that

$$\{s > 0\} \subseteq E_i + N$$

and

$$\nu(E_j^c) = 0 \quad \text{where } j = i + m_0 \pmod{m}.$$

Proof. Suppose that j is such that $\nu(E_j) > 0$. Then for any $x \in E$, $s(x) > 0$ implies

$$K^{m_0}(x, E_j) \geq \beta s(x) \nu(E_j) > 0.$$

This is possible only if $x \in E_{j-m_0}$ or $x \in N$. There cannot be any other index $j' \neq j$ with $\nu(E_{j'}) > 0$, since this would contradict the hypotheses $\psi(s) > 0$ and $\psi(N) = 0$. □

By Remark 2.1(ii) there is no loss of generality in assuming that $\nu(s) > 0$. Let d be the greatest common divisor (abbreviated g.c.d.) of the set

$$I = \{m \geq 1 : M(m, \beta_m, s, \nu) \text{ for some } \beta_m > 0\}. \tag{2.6}$$

It is clear that I is closed under addition and then it contains all sufficiently large multiples of d.

The following theorem solves the problem of the existence and uniqueness of cycles:

Theorem 2.2. Suppose that K is an irreducible kernel. Let $d \geq 1, s \in \mathcal{S}^+$ and $\nu \in \mathcal{M}^+$, $\nu(s) > 0$, be as above. Then:

(i) There is a d-cycle $(E_0, \ldots, E_{d-1})$. The integer d does not depend on the particular choice of the small function s and of the small measure ν.

(ii) If $(E'_0, \ldots, E'_{d'-1})$ is another cycle, then d' divides d, and any E'_i is the union (ψ-a.e.) of d/d' sets from the collection $\{E_0, \ldots, E_{d-1}\}$. In particular, if $(E'_0, \ldots, E'_{d-1})$ is another d-cycle, then $E'_i = E_j$ (ψ-a.e.), where $j = i + r$ (mod d), for some integer $0 \leq r < d$.

Proof. For any $i = 0, \ldots, d-1$, set

$$\tilde{E}_i = \left\{ \sum_{n=1}^{\infty} K^{nd-i} s > 0 \right\}.$$

By irreducibility, $\tilde{E}_0 \cup \ldots \cup \tilde{E}_{d-1} = E$. Should $\tilde{E}_i \cap \tilde{E}_j \in \mathcal{E}^+$ hold for some $i \neq j$, then $\{K^{nd-i}s > 0\} \cap \{K^{n'd-j}s > 0\} \in \mathcal{E}^+$ for some $n, n' \geq 1$, and by irreducibility, $\nu K^{q+nd-i}s > 0$ and $\nu K^{q+n'd-j}s > 0$ for some $q \geq 0$. But then both $q + nd - i + 2m_0$ and $q + n'd - j + 2m_0$ would belong to the set I, which contradicts the definition of the integer d.

By Proposition 2.5 there is a closed (and hence full) set F such that the sets $E_i = \tilde{E}_i \cap F$ are disjoint. Their union $\sum_0^{d-1} E_i$ is equal to F. Now, if $x \in F$ is such that $K(x, E_j) > 0$ then clearly x belongs to E_i, for $i = j - 1$ (mod d). This proves that $(E_0, \ldots, E_{d-1})$ is a d-cycle.

The uniqueness assertion (ii) follows from Proposition 2.8. The independence of d of the choice of s and ν is a direct consequence of (ii). □

In what follows we shall assume that $(E_0, \ldots, E_{d-1})$ is a fixed d-cycle. The sets $E_0, \ldots, E_{d-1}$ are called *cyclic sets*. N denotes the ψ-null set $N = (E_0 + \cdots + E_{d-1})^c$. The integer $d \geq 1$ is called the *period* of the kernel K. If $d = 1$, K is called *aperiodic*, otherwise *periodic*. By convention, in the sequel addition and equalities involving indices of the cyclic sets are always modulo d.

Examples 2.5. (*a*) Suppose that K is a Card_F-irreducible matrix. Fix any state $x_0 \in F$. Then (with obvious notation)

$$d = \text{g.c.d.}\{m \geq 1 : k^m(x_0, x_0) > 0\}$$

is the period of K. The cyclic sets are given by

$$E_i = \{x \in F : k^{nd-i}(x, x_0) > 0 \quad \text{for some } n \geq 1\}, \quad i = 0, \ldots, d-1.$$

(*c*) Suppose that F is spread-out. Then the random walk (Z_n) is aperiodic.

(*d*) If $F(\mathbb{R}_+) < 1$, the reflected random walk (W_n) is aperiodic.

(*e*) Suppose that F is spread-out. Then the Markov chains $(V^+_{n\delta}; n \geq 0)$, $\delta > 0$, are aperiodic.

(*f*) With the assumptions of Example 2.1 (*f*), the autoregressive process (R_n) is aperiodic.

For later purposes we shall discuss the cyclic behaviour of the iterates K^m of K, $m \geq 1$.

By the remarks after Definition 2.4, K^d 'splits' into d distinct kernels with respective state spaces $(E_i, \mathscr{E} \cap E_i)$, $i = 0, \ldots, d-1$. In fact, it follows from Definition 2.4 and Theorem 2.2 that, for an arbitrary integer $m \geq 1, K^m$ splits into $c_m = \text{g.c.d.}\{m, d\}$ distinct kernels having state spaces

$$E_i^{(m)} = E_i + E_{i+c_m} + E_{i+2c_m} + \cdots + E_{i+d-c_m}, \quad i = 0, 1, \ldots, c_m - 1,$$

respectively.

As an immediate consequence of the definitions we have the following result:

Proposition 2.9. Let $m \geq 1$ be arbitrary and let $0 \leq i \leq c_m - 1$. Then the kernel K^m with state space $(E_i^{(m)}, \mathscr{E} \cap E_i^{(m)})$ is $\psi I_{E_i^{(m)}}$-irreducible and has period d/c_m.

In particular, the kernel K^d with state space E_i is ψI_{E_i}-irreducible and aperiodic for each $i = 0, 1, \ldots, d-1$.

Proof. Obvious. □

For the rest of this section we return to the study of the class $\mathscr{S}^+$ of small functions.

For $i = 0, \ldots, d-1$, let $\mathscr{S}_i^+$ denote the class of those small functions s which vanish on the cyclic sets $E_j, j \neq i$. Recall from Proposition 2.8 that every small function belongs to one of the subclasses $\mathscr{S}_i^+$:

$$\mathscr{S}^+ = \sum_0^{d-1} \mathscr{S}_i^+. \tag{2.7}$$

Proposition 2.10. For every $i = 0, \ldots, d-1$:

(i) If $s \in \mathscr{S}_i^+$ then $Ks \in \mathscr{S}_{i-1}^+$.

(ii) If $s \in \mathscr{S}_i^+$, and $s' \in \mathscr{E}^+$ is such that $s' \leq \gamma s$ for some constant $\gamma > 0$, then also $s' \in \mathscr{S}_i^+$.

(iii) If s and $s' \in \mathscr{S}_i^+$ then $s + s' \in \mathscr{S}_i^+$ and $s \vee s' \in \mathscr{S}_i^+$.

(iv) If $s \in \mathscr{S}_i^+$ then $\sum_{n=0}^m K^{nd+q} s \in \mathscr{S}_{i-q}^+$, for all $m, q \geq 0$.

Proof. (i) and (ii) are obvious. (iv) is an immediate consequence of (i) and (iii).

In order to prove (iii) take two arbitrary small functions $s, s' \in \mathscr{S}_i^+$. By Remarks 2.1(ii) and (iii) there is no loss of generality in assuming that we have $M(m_0, 1, s, \nu)$ and $M(m_0', 1, s', \nu')$ with $\nu(s) > 0$ and $\nu'(s') > 0$ for some m_0, m_0', ν and ν'. Since the set I given by (2.6) contains all sufficiently great multiples of d, we can find integers m and m' such that $\nu K^m s > 0$, $\nu' K^{m'} s > 0$

and $m = m' + m_0' - m_0$. It follows that

$$K^{m+2m_0} \geq (\nu K^m s)s \otimes \nu$$

and

$$K^{m+2m_0} = K^{m_0'+m'+m_0} \geq (\nu' K^{m'} s)s' \otimes \nu,$$

and hence the function $\frac{1}{2}(\nu K^m s)s + \frac{1}{2}(\nu' K^{m'} s)s'$ is small. The final assertion (iii) follows easily now from (ii). □

Similar statements are valid for small measures. We leave the details to the reader.

For every $i = 0, \ldots, d-1$, denote by $\mathscr{S}_i^+$ the subclass of small sets $C \in \mathscr{S}^+$ such that $1_C \in \mathscr{S}_i^+$, i.e., $C \cap E_j = \varnothing$ for all $j \neq i$. By (2.7) every small set $C \in \mathscr{S}^+$ belongs to one of the subclasses $\mathscr{S}_i^+$.

The most important result in the next proposition is part (iv) stating that the state space E can be decomposed into a countable number of small sets.

Proposition 2.11. (i) Let $s \in \mathscr{S}_i^+$ be small. Every set $C \in \mathscr{E}^+$ satisfying

$$\inf_{x \in C} \sum_{n=0}^{m} K^{nd+q} s(x) > 0 \quad \text{for some } m, q \geq 0,$$

is small and belongs to the subclass $\mathscr{S}_{i-q}^+$.

(ii) Any set $C \in \mathscr{E}^+$ satisfying the following condition is small: there is a set $B \in \mathscr{E}^+$ such that for all $A \in \mathscr{E}^+$ with $A \subseteq B$,

$$\inf_{x \in C} \sum_{n=0}^{m} K^{nd+q}(x, A) > 0 \quad \text{for some } m = m(A),\ q = q(A) \geq 0.$$

(iii) The subclass $\mathscr{S}_i^+$ of small sets is closed under finite unions.

(iv) There is a countable partition $E = \sum_{m=1}^{\infty} C_m$ of $(E, \mathscr{E})$ into small sets $C_m \in \mathscr{S}^+$.

Proof. (i) and (iii) are immediate consequences of Proposition 2.10. (ii) follows from (i) and Proposition 2.6. For the proof of (iv) set

$$C_m^{(i)} = \left\{ \sum_{n=1}^{m} K^{nd-i} s \geq m^{-1} \right\}.$$

By (i), each $C_m^{(i)}$ is small. Using the same notation as in the proof of Theorem 2.2 we see that their union, $\lim_{m \to \infty} \uparrow C_m^{(i)}$ is equal to $\tilde{E}_i$. We have $\bigcup_{i=0}^{d-1} \tilde{E}_i = E$, and hence the result follows. □

In the sequel, in order to avoid unessential technicalities, we shall mostly concentrate on the aperiodic case. For convenience we write out Proposition 2.11 in this special case:

Corollary 2.1. Let K be aperiodic. Then:

(i) The functions $\sum_0^m K^n s$, $m \geq 0$, are small. In particular any set $C \in \mathscr{E}^+$ satisfying

$$\inf_{x \in C} \sum_0^m K^n s(x) > 0 \quad \text{for some } m \geq 0$$

is small.

(ii) Any set $C \in \mathscr{E}^+$ satisfying the following condition is small: There is a set $B \in \mathscr{E}^+$ such that for all $A \in \mathscr{E}^+$ with $A \subseteq B$,

$$\inf_{x \in C} \sum_0^m K^n(x, A) > 0 \quad \text{for some } m = m(A) \geq 0.$$

(iii) The class $\mathscr{S}^+$ of small sets is closed under finite unions.

(iv) There exists an increasing sequence $C_1 \subseteq C_2 \subseteq \ldots$ of small sets $C_m \in \mathscr{S}^+$ such that $\lim_{m \to \infty} \uparrow C_m = E$. □

3

Transience and recurrence

Our aim in this chapter is to define and investigate the concepts of transience and recurrence.

First we consider a general irreducible kernel K, showing that there exists a constant $0 \leq R < \infty$ (to be called the convergence parameter of K), the powers R^{-n} of which describe the growth of the iterates K^n as $n \to \infty$. Roughly speaking, the potential kernel $G^{(r)} = \sum_0^\infty r^n K^n$ of the kernel rK is 'finite' for $r < R$ and 'infinite' for $r > R$. If $G^{(r)}$ is 'finite' also at the critical value $r = R$, we say that the kernel K is R-transient; otherwise K is R-recurrent.

After these general results we will concentrate on the case where $K = P$ is the (not necessarily irreducible) transition probability of a Markov chain (X_n). The main result here is Hopf's decomposition theorem stating that the state space $(E, \mathscr{E})$ can be divided into two parts, E_{d} (called the dissipative part) and E_{c} (the conservative part). The dissipative part E_{d} is a countable union of sets B_i, such that the Markov chain (X_n) is transient on each B_i; i.e.

$$\mathbb{P}_x\{X_n \in B_i \quad \text{i.o.}\} = 0 \quad \text{for all } x \in E$$

('i.o.' means 'infinitely often'). On the conservative part E_{c} the Markov chain is recurrent in the following sense. There is a non-trivial σ-finite measure φ on $(E, \mathscr{E})$ such that for any φ-positive set $B \subseteq E_{\mathrm{c}}$, for φ-almost all $x \in B$:

$$\mathbb{P}_x\{X_n \in B \quad \text{i.o.}\} = 1. \tag{3.1}$$

When (X_n) is irreducible, the concepts of recurrence and 1-recurrence coincide, and then there is even an absorbing set H such that (3.1) holds for all $B \in \mathscr{E}^+, x \in H$. In this case we say that the Markov chain (X_n) is Harris recurrent on H.

3.1 Some potential theory

Since it will be useful to have some potential theoretic notions available, we start with a brief excursion into the potential theory of non-negative kernels.

Recall from Section 2.1 the definition of the potential kernel $G = \sum_0^\infty K^n$.

Definition 3.1. A non-negative function $h \in \mathscr{E}_+$, which is not identically infinite, is called *superharmonic* (resp. *harmonic*) *for the kernel* K, if

$$h \geq Kh \quad (\text{resp. } h = Kh).$$

A function $p \in \mathscr{E}_+$, $p \not\equiv \infty$, is called a *potential*, if there is a function $g \in \mathscr{E}_+$, called the *charge* of p, such that

$$p = Gg.$$

Proposition 3.1. Every potential is superharmonic.

Proof. By Proposition 2.1 we have $p = g + Kp \geq Kp$. □

Proposition 3.2. Suppose that h is superharmonic. Then:

(i) The set $\{h < \infty\}$ is closed.

(ii) Either $h > 0$ everywhere or the set $\{h = 0\}$ is closed.

(iii) If K is irreducible and $h \in \mathscr{E}^+$, then in fact $h > 0$ everywhere.

(iv) If K is substochastic and $0 \leq h \leq 1$ is harmonic, then either $h < 1$ everywhere or the set $\{h = 1\}$ is absorbing.

Proof. (i) When $x \in \{h < \infty\}$ we have $\infty > h(x) \geq Kh(x) \geq \infty \cdot K(x, \{h = \infty\})$. Therefore $K(x, \{h = \infty\}) = 0$.

(ii) and (iv): The proofs are similar to that of (i).

(iii) This is a direct consequence of (ii) and Proposition 2.5(i). □

The fundamental result in potential theory is the *Riesz decomposition theorem*:

Theorem 3.1. (i) If $h \in \mathscr{E}_+, h \not\equiv \infty$, is the sum of a potential $p = Gg$ and a harmonic function h^∞,

$$h = p + h^\infty, \tag{3.2}$$

then h is superharmonic, and on the closed set $F = \{h < \infty\}$ we have

$$g = h - Kh \tag{3.3}$$

and

$$h^\infty = \lim_{h \to \infty} \downarrow K^n h. \tag{3.4}$$

(ii) Conversely, if h is superharmonic, then on $F = \{h < \infty\}$, h is the sum of a potential $p = Gg$ and a harmonic function h^∞, where g is given by (3.3) and h^∞ by (3.4).

(iii) If $h \in \mathscr{E}_+$ is superharmonic, and $g \in \mathscr{E}_+$ is such that

$$h \geq g + Kh,$$

then

$$h \geq Gg.$$

Proof. (i) and (ii): If (3.2) holds, then by Proposition 2.1, and since h^∞ is harmonic

$$h = g + Kp + Kh^{\infty} = g + Kh.$$

Thus h is superharmonic and (3.3) holds on $\{h < \infty\}$.

Suppose now that h is superharmonic, and let $g \in \mathscr{E}_+$ be such that $h = g + Kh$. By iterating this equation we obtain

$$h = \sum_0^{n-1} K^m g + K^n h \quad \text{for all } n \geq 1.$$

As $n \to \infty$, the first term on the right hand side increases to $Gg = p$; the second decreases to a function h^{∞}, which is harmonic on $\{h < \infty\}$ by the monotone convergence theorem.

(iii) The proof follows again by iterating. □

The pair (p, h^{∞}) appearing in Theorem 3.1 is called the *Riesz decomposition of the superharmonic function h* (on $F = \{h < \infty\}$).

We call part (iii) the *balayage theorem.*

The following simple result will be needed in the sequel:

Proposition 3.3. Suppose that h is a finite superharmonic function. Let $h^{\infty} = \lim\downarrow K^n h$. Then either $h = p = Gg$ is a potential (and $h^{\infty} \equiv 0$), or $\sup_E (h^{\infty}/h) = 1$.

Proof. In any case $0 \leq h^{\infty} \leq h$. If $h^{\infty} \leq \rho h$ for some constant $0 \leq \rho < 1$, then

$$h^{\infty} = K^n h^{\infty} \leq \rho K^n h \downarrow \rho h^{\infty} \quad \text{as } n \to \infty,$$

which is possible only if $h^{\infty} \equiv 0$. □

3.2 *R*-transience and *R*-recurrence

Suppose that the kernel K is irreducible. In this section we shall define and study concepts related to the 'rate of growth' of the iterates K^n of K.

We adopt the notation used in Chapter 2 for irreducible K; in particular, ψ denotes a maximal irreducibility measure, $\mathscr{E}^+ = \{f \in \mathscr{E}_+ : \psi(f) > 0\}$ and $\mathscr{S}^+ = \{s \in \mathscr{E}^+ : s \text{ small}\}$. Recall from Proposition 2.11(iv) that there is a countable partition $E = \sum_{m=1}^{\infty} C_m$ of $(E, \mathscr{E})$ into small sets $C_m \in \mathscr{S}^+$. Recall also that every closed set F is full; this means $\psi(F^c) = 0$.

For any $0 \leq r < \infty$, denote by $G^{(r)}$ the potential kernel of the kernel rK, i.e. $G^{(r)} = \sum_0^{\infty} r^n K^n$.

Definition 3.2. A real number $0 \leq R < \infty$ is called the *convergence parameter of the kernel K*, provided that there exists a closed set F such that

(i) $\quad G^{(r)} s < \infty$ on F, for all $0 \leq r < R$, all $s \in \mathscr{S}^+$, and

(ii) $\quad G^{(r)}f \equiv \infty \quad$ for all $r > R$, all $f \in \mathscr{E}^+$.

The kernel K is called *R-transient* (resp. *R-recurrent*), if (i) (resp. (ii)) holds when $r = R$.

It is not immediately clear that there should exist any convergence parameter, or that K should be either R-transient or R-recurrent. However, we have the following:

Theorem 3.2. Suppose that the kernel K is irreducible. Then:

(i) There exists a convergence parameter $0 \leq R < \infty$.

(ii) The kernel K is either R-transient or R-recurrent.

Proof. Let us fix a small function $s \in \mathscr{S}^+$. Set

$$R = \sup\{r \geq 0 : G^{(r)}s(x) < \infty \text{ for some } x \in E\},$$
$$F^{(r)} = \{G^{(r)}s < \infty\}, \quad 0 \leq r < \infty.$$

Clearly, for all $0 \leq r < R$, the function $G^{(r)}s$ is a potential, whence a superharmonic function, for the kernel rK. By Proposition 3.2(i) the sets $F^{(r)}, 0 \leq r < R$, are closed, and hence, by Proposition 2.5(iii), so is their intersection

$$F = \bigcap_{0 \leq r < R} F^{(r)} = \lim_{r_m \uparrow R} \downarrow F^{(r_m)}.$$

(In the special case when $R = 0$, we set $F = F^{(0)} = E$.)

Let $s' \in \mathscr{S}^+$ be another small function. It follows from Proposition 2.7(ii) that $G^{(r)}s' < \infty$ on F, for all $0 \leq r < R$. By the same proposition we also see that $G^{(r)}f \equiv \infty$ for all $r > R$, $f \in \mathscr{E}^+$. Similarly we find that K is either R-transient or R-recurrent.

It remains to prove that the convergence parameter R is finite. To this end, suppose that the minorization condition $M(m_0, \beta, s, \nu)$ holds with $\nu(s) > 0$ (cf. Remark 2.1(ii)). Then

$$K^{nm_0}s \geq \beta^n (s \otimes \nu)^n s = (\beta \nu(s))^n s.$$

It follows that $G^{(r)}s \geq \sum_{n=0}^{\infty} r^{nm_0} K^{nm_0} s = \infty$ on the ψ-positive set $\{s > 0\}$, for all $r \geq (\beta\nu(s))^{-1/m_0}$. According to what we have proved before this implies that $R \leq (\beta\nu(s))^{-1/m_0}$. □

In what follows, if we say that K is R-transient (or R-recurrent) then, by convention, this implicitly states that the convergence parameter of K is R.

Examples 3.1. (*a*) Let K be a Card_F-irreducible matrix, Card_F being a maximal irreducibility measure (cf. Example 2.2(*a*)). Let

$$g^{(r)}(x, y) = \sum_{n=0}^{\infty} r^n k^n(x, y), \quad 0 \le r < \infty.$$

We have

$$g^{(r)}(x, y) < \infty \quad \text{for all } 0 \le r < R,\ x \in F,\ y \in E,$$

and

$$g^{(r)}(x, y) = \infty \quad \text{for all } r > R,\ x \in E,\ y \in F;$$

for $r = R$, the former or latter statement holds true, depending on whether K is R-transient or R-recurrent.

(*b*) Let K be an irreducible integral kernel with basis φ and density k, and with maximal irreducibility measure $\varphi = \varphi I_{B'}$ (cf. Example 2.2(*b*)). There exists a closed set $F \subseteq B', \varphi(B' \backslash F) = 0$, such that

$$g^{(r)}(x, y) \stackrel{\text{def}}{=} \sum_{n=1}^{\infty} r^n k^{(n)}(x, y) < \infty$$

$$\text{for all } 0 \le r < R,\ x \in F,\ \varphi\text{-almost all } y \in E,$$

and

$$g^{(r)}(x, y) = \infty \quad \text{for all } r > R,\ x \in E,\ \varphi\text{-almost all } y \in F;$$

for $r = R$, the former or latter statement holds true, depending on whether K is R-transient or R-recurrent.

We introduce a new example:

(*h*) Let $(Z_n; n \ge 0)$ be a *multitype branching process* with general type space $(E, \mathscr{E})$; i.e., we assume that, for each $n \ge 0$, an individual in the nth generation of the type x produces a random number of children with random types independently of the earlier generations $0, 1, \ldots, n-1$ and of the other individuals in the nth generation. If we denote by $Z_n(A)$ the number of individuals belonging to the set $A\,(\in \mathscr{E})$ in the nth generation, and set

$$M(x, A) = \mathbb{E}[Z_1(A) \mid Z_0 = \varepsilon_x] \quad (\varepsilon_x(B) \stackrel{\text{def}}{=} 1_B(x),\ x \in E,\ B \in \mathscr{E}),$$

it follows that the nth iterate M^n of the kernel M has the interpretation

$$M^n(x, A) = \mathbb{E}[Z_n(A) \mid Z_0 = \varepsilon_x].$$

The convergence parameter R of the kernel M is called the *Malthusian parameter* of the branching process $(Z_n; n \ge 0)$. (See e.g. Harris, 1963.)

Remarks 3.1. (i) A sufficient (and necessary) condition for $R > 0$ is: For some $x_0 \in E, f_0 \in \mathscr{E}^+, M_0 < \infty, \gamma_0 < \infty$, we have

$$K^n f_0(x_0) \le M_0 \gamma_0^n \quad \text{for all } n \ge 0.$$

Then in fact, $R \ge \gamma_0^{-1}$.

(ii) If $K = P$ is a transition probability then $1 \leq R < \infty$. As a corollary of the above theorem we have in this case: There is a closed set F such that either $Gs < \infty$ on F for all $s \in \mathscr{S}^+$, or $Gf \equiv \infty$ for all $f \in \mathscr{E}^+$. Later in Corollary 3.1 we will see that in the former case even $\sup_E Gs < \infty$, for all $s \in \mathscr{S}^+$.

Later we will need the following:

Proposition 3.4. For any small measure $\nu \in \mathscr{M}^+$ and small function $s \in \mathscr{S}^+$:

(i) $\nu G^{(r)} s < \infty$ for $0 \leq r < R$ and $\nu G^{(r)} s = \infty$ for $r > R$.

(ii) $\nu G^{(R)} s = \infty$ if and only if K is R-recurrent.

Proof. Use Proposition 2.7(iii) and Theorem 3.2. □

For later purposes we consider next the 'R-properties' of the iterated kernels K^m, $m \geq 2$. Recall from Proposition 2.9 that there are in fact $c_m (= \text{g.c.d.}\{m, d\})$ distinct kernels K^m with respective state spaces $E_i^{(m)}$, $0 \leq i < c_m$, all of which are irreducible with period d/c_m. Define $d'_m = md/c_m$.

Proposition 3.5. If the kernel K is R-transient (resp. R-recurrent) then all the c_m kernels K^m are R^m-transient (resp. R^m-recurrent).

Proof. There is no loss of generality in considering K^m on the state space $E_0^{(m)} = E_0 + E_{c_m} + \cdots + E_{d-c_m}$. We adopt the notation used in the proof of Theorem 2.2. In particular, m_0, s and ν denote an integer, a small function and a small measure, respectively, such that $M(m_0, 1, s, \nu)$ and $\nu(s) > 0$ hold. Note that by the construction made in the proof of Theorem 2.2 and by Proposition 2.8, the set $E_0 + N$ supports both s and ν.

Write $R^{(m)}$ for the convergence parameter of K^m. Clearly $r < R$ implies that $r^m \leq R^{(m)}$. Hence $R^{(m)} \geq R^m$.

Suppose now that $r < (R^{(m)})^{1/m}$. Then the series $\sum_{n=0}^{\infty} r^{nm} K^{nm} s$ converges ψ-a.e. on E_0. Let $n_0 \geq 1$ be an integer and $\gamma > 0$ a constant such that

$$r^{n_0 d_m - jd} K^{n_0 d_m - jd} s \geq \gamma s \quad \text{for all } j = 0, 1, \ldots, \frac{m}{c_m} - 1.$$

(This is possible since the set I given by (2.6) contains all sufficiently great multiples of d.) It follows that, ψ-a.e. on E_0:

$$\infty > \frac{m}{c_m} \sum_{n=0}^{\infty} r^{(n+n_0)d_m} K^{(n+n_0)d_m} s$$

$$\geq \gamma \sum_{n=0}^{\infty} \sum_{j=0}^{m/c_m - 1} r^{nd_m + jd} K^{nd_m + jd} s$$

$$= \gamma \sum_{n=0}^{\infty} r^{nd} K^{nd} s.$$

Since, by Proposition 2.10(i), $K^{nd+i}s$ vanishes on E_0 for all $n \geq 0, i = 1, \ldots, d-1$, we see that $G^{(r)}s$ is finite ψ-a.e. on E_0. Therefore $r < R$. This proves the converse inequality $R^{(m)} \leq R^m$.

Setting $r = R$ in the above proof we also see that K^m is R^m-transient if and only if K is R-transient. □

3.3 Stopping times for Markov chains

Throughout Sections 3.3–3.6 *we assume that* $K = P$ *is the* (not necessarily irreducible) *transition probability of a Markov chain* $(X_n; n \geq 0)$. In Sections 3.4–3.6 we will need the concept of a stopping time and the associated notion of strong Markov property. For later purposes (see Section 4.4) we introduce in this context also the concept of a randomized stopping time.

Let $(\mathscr{F}_n)$ be a history. An $\bar{\mathbb{N}}$-valued random variable T is called a *random time*. If

$$\{T=n\} \in \mathscr{F}_n \quad \text{for all } n \geq 0,$$

then the random time T is called a *stopping time relative to the history* $(\mathscr{F}_n)$. If in addition, (X_n) is Markov w.r.t. $(\mathscr{F}_n)$, then T is called a *stopping time for the Markov chain* $(X_n, \mathscr{F}_n)$. If T is a stopping time relative to the internal history $(\mathscr{F}_n^X)$ then it is called simply a *stopping time for the Markov chain* (X_n). This means that, for all $n \geq 0$,

$$\{T=n\} = \{(X_0, \ldots, X_n) \in B_n\}$$

for some set $B_n \in \mathscr{E}^{\otimes (n+1)}$.

An important example of a stopping time for the Markov chain (X_n) is the hitting time T_B of a set $B \in \mathscr{E}$, defined by $T_B = \inf\{n \geq 0 : X_n \in B\}$.

A random time T is called a *randomized stopping time* (for the Markov chain (X_n)), if, for every $n \geq 0$, the event $\{T=n\}$ and the post-n-chain $(X_n, X_{n+1}, \ldots)$ are conditionally independent, given the pre-n-chain $(X_0, \ldots, X_n)$, i.e.

$$\mathbb{P}\{T=n \mid \mathscr{F}^X\} = \mathbb{P}\{T=n \mid \mathscr{F}_n^X\} = f_n(X_0, \ldots, X_n)$$

for some $\mathscr{E}^{\otimes (n+1)}$-measurable function f_n. We set

$$\mathbb{P}\{T=\infty \mid \mathscr{F}^X\} = f_\infty(X_0, X_1, \ldots) = 1 - \sum_0^\infty f_n(X_0, \ldots, X_n).$$

Clearly a random time T is a randomized stopping time, if and only if for every $n \geq 0$,

$$\begin{aligned}\mathbb{P}\{T=n \mid \mathscr{F}^X; T \geq n\} &= \mathbb{P}\{T=n \mid \mathscr{F}_n^X; T \geq n\} \\ &= r_n(X_0, \ldots, X_n)\end{aligned}$$

for some $\mathscr{E}^{\otimes (n+1)}$-measurable function r_n. The functions f_n and r_n are related to each other through the formulas

$$r_n = \left(1 - \sum_0^{n-1} f_m\right)^{-1} f_n,$$

$$f_n = (1 - r_0)\ldots(1 - r_{n-1})r_n,$$

$$f_\infty = (1 - r_0)(1 - r_1)\ldots.$$

By convention, we regard two randomized stopping times T and T' as equal, and write

$$T \overset{\mathscr{L}}{=} T',$$

provided that their conditional distributions, given $\mathscr{F}^X$, coincide.

As an example of a randomized stopping time take any $0 \leq g \leq 1$, and set $r_n(x_0, \ldots, x_n) = g(x_n), n \geq 0$. The resulting randomized stopping time, denoted by T_g, can be regarded as a geometric random variable where the probability of success at epoch n depends on the current state X_n of the Markov chain through the function g. Note that, if $g = 1_B$ is the indicator of a set $B \in \mathscr{E}$, we have $T_g = T_{1_B} \overset{\mathscr{L}}{=} T_B$.

A stopping time for the Markov chain (X_n) is clearly a randomized stopping time. Conversely, a randomized stopping time for (X_n) is a stopping time for $(X_n, \mathscr{F}_n)$, for a conveniently chosen history $(\mathscr{F}_n)$. In fact the following holds:

Proposition 3.6. (i) Suppose that (X_n) is Markov w.r.t. a history $(\mathscr{F}_n)$. If T is a stopping time for the Markov chain $(X_n, \mathscr{F}_n)$, then it is also a randomized stopping time for (X_n).

(ii) Conversely, if T is a randomized stopping time for (X_n), then there is a history $(\mathscr{F}_n)$ such that (X_n) is Markov w.r.t. $(\mathscr{F}_n)$ and T is a stopping time for the Markov chain $(X_n, \mathscr{F}_n)$. One can choose $\mathscr{F}_n = \mathscr{F}_n^X \vee \mathscr{F}_n^T$, where $\mathscr{F}_n^T \overset{\text{def}}{=} \sigma(T = m; 0 \leq m \leq n)$, $n \geq 0$.

Proof. Let ζ be an arbitrary non-negative functional.

(i) Let T be a stopping time for the Markov chain $(X_n, \mathscr{F}_n)$. By the hypotheses and Theorem 1.1,

$$\begin{aligned} \mathbb{E}[\zeta \circ \theta_n; T = n | \mathscr{F}_n^X] &= \mathbb{E}[\zeta \circ \theta_n; T = n | \mathscr{F}_n | \mathscr{F}_n^X] \\ &= \mathbb{E}[\mathbb{E}[\zeta \circ \theta_n | \mathscr{F}_n]; T = n | \mathscr{F}_n^X] \\ &= \mathbb{E}_{X_n}[\zeta] \mathbb{P}\{T = n | \mathscr{F}_n^X\}, \end{aligned}$$

which proves the desired conditional independence.

(ii) Let T be a randomized stopping time. We have for any $0 \leq m \leq n$:

$$\mathbb{E}[\zeta\circ\theta_n; T=m|\mathscr{F}_n^X] = \mathbb{E}[\zeta\circ\theta_n \mathbb{P}\{T=m|\mathscr{F}^X\}|\mathscr{F}_n^X]$$
$$= \mathbb{E}_{X_n}[\zeta]\mathbb{P}\{T=m|\mathscr{F}_n^X\},$$

so that

$$\mathbb{E}[\zeta\circ\theta_n|\mathscr{F}_n^X \vee \mathscr{F}_n^T] = \mathbb{E}_{X_n}[\zeta].$$

In other words, (X_n) is Markov w.r.t. the history $(\mathscr{F}_n) = (\mathscr{F}_n^X \vee \mathscr{F}_n^T)$. Trivially, T is a stopping time relative to $(\mathscr{F}_n)$. □

Theorem 1.1 expresses the Markov property of (X_n) at the deterministic time epochs n. According to the following theorem (X_n) is *strong Markov*, that is, it has the Markov property also at stopping times (or what by the preceding proposition is equivalent, at randomized stopping times).

Suppose that T is a stopping time relative to a history $(\mathscr{F}_n)$. We can associate with T the σ-algebra $\mathscr{F}_T$ defined by

$$\mathscr{F}_T = \{A \in \bigvee_0^\infty \mathscr{F}_n:\ A \cap \{T \leq n\} \in \mathscr{F}_n \quad \text{for all } n \geq 0\},$$

the σ-algebra of events which 'happen before T'.

Theorem 3.3. Suppose that (X_n) is Markov w.r.t. a history $(\mathscr{F}_n)$, and let T be an arbitrary randomized stopping time. Then for any non-negative functional ζ,

$$\mathbb{E}[\zeta\circ\theta_n|\mathscr{F}_n; T=m] = \mathbb{E}_{X_n}[\zeta] \quad \text{for all } 0 \leq m \leq n.$$

In particular, if T is a stopping time for the Markov chain $(X_n, \mathscr{F}_n)$ then the following strong Markov property holds:

$$\mathbb{E}[\zeta\circ\theta_T|\mathscr{F}_T] = \mathbb{E}_{X_T}[\zeta] \quad \text{on } \{T < \infty\}.$$

Proof. It follows from Proposition 3.6 that the Markov chain $(X_n, \mathscr{F}_n)$ is Markov also w.r.t. the history $(\mathscr{F}_n^X \vee \mathscr{F}_n^T)$. Consequently, it is Markov w.r.t. $(\mathscr{F}_n \vee \mathscr{F}_n^X \vee \mathscr{F}_n^T) = (\mathscr{F}_n \vee \mathscr{F}_n^T)$. The first assertion then reduces to the ordinary Markov property w.r.t. the history $(\mathscr{F}_n \vee \mathscr{F}_n^T)$.

The strong Markov property now follows from the formula

$$\mathbb{E}[\zeta\circ\theta_T|\mathscr{F}_T] = \mathbb{E}[\zeta\circ\theta_n|\mathscr{F}_n] \quad \text{on } \{T=n\},$$

and from Theorem 1.1. □

3.4 Hitting and exit times

Let $B \in \mathscr{E}$ be arbitrary. We denote by T_B (resp. S_B) the *first hitting time* of the set B, including time 0 (resp. not including time 0):

$$T_B = \inf\{n \geq 0: X_n \in B\},$$
$$S_B = \inf\{n \geq 1: X_n \in B\} = T_B \circ \theta.$$

(By convention, $\inf \varnothing = \infty$.) The iterates $T_B(i)$, $i \geq 0$, are defined by $T_B(0) = T_B$, and

$$T_B(i) = \inf\{n > T_B(i-1): X_n \in B\}$$
$$= m + S_B \circ \theta_m \quad \text{on } \{T_B(i-1) = m\}, i \geq 1.$$

Similarly, the iterates $S_B(i), i \geq 1$, are defined by $S_B(1) = S_B$, and

$$S_B(i) = \inf\{n > S_B(i-1): X_n \in B\}$$
$$= m + S_B \circ \theta_m \quad \text{on } \{S_B(i-1) = m\}, i \geq 2.$$

So, for example, $T_B(i)$ is the epoch of the $(i+1)$th visit to the set B. Clearly all the random times $T_B(i)$ and $S_B(i)$ are stopping times for the Markov chain (X_n).

Let us denote by I_B the *multiplication by* 1_B, i.e., I_B is the kernel

$$I_B(x, A) = 1_{A \cap B}(x), \quad x \in E, A \in \mathscr{E}.$$

Note that $I_E = I$, the identity kernel.

Let us define the kernels G_B, G'_B and U_B by

$$G_B(x, A) = \mathbb{E}_x \sum_0^{T_B} 1_A(X_n) = \sum_0^\infty \mathbb{P}_x\{X_n \in A, T_B \geq n\}$$

$$= \sum_0^\infty (I_{B^c} P)^n(x, A), \tag{3.5}$$

$$G'_B(x, A) = \mathbb{E}_x \sum_0^{S_B - 1} 1_A(X_n) = \sum_0^\infty \mathbb{P}_x\{X_n \in A, S_B > n\}$$

$$= \sum_0^\infty (P I_{B^c})^n(x, A) \tag{3.6}$$

$$U_B(x, A) = \mathbb{E}_x \sum_1^{S_B} 1_A(X_n) = \sum_0^\infty P(I_{B^c} P)^n(x, A), \quad x \in E, A \in \mathscr{E}. \tag{3.7}$$

Note that G_B (resp. G'_B) is the potential kernel of the kernel $I_{B^c} P$ (resp. $P I_{B^c}$), and that

$$U_B = P G_B = G'_B P, \tag{3.8}$$

$$G_B = I + I_{B^c} U_B, \tag{3.9}$$

$$G'_B = I + U_B I_{B^c}. \tag{3.10}$$

Also note the interpretations

$$G_B I_B(x, A) = \mathbb{P}_x\{X_{T_B} \in A, T_B < \infty\}, \tag{3.11}$$

$$U_B I_B(x, A) = \mathbb{P}_x\{X_{S_B} \in A, S_B < \infty\}, \tag{3.12}$$

and the special cases $G_\varnothing = G'_\varnothing = G, G_E = G'_E = I, U_E = P$.

Proposition 3.7. (i) If $A \subseteq B$, then

$$G_A = G_B I_{B^c} + G_B I_B G_A, \tag{3.13}$$

$$U_A = U_B + U_B I_{B \setminus A} U_A. \tag{3.14}$$

In particular,

$$G = G_B I_{B^c} + G_B I_B G, \tag{3.15}$$

and for any $f \in \mathscr{E}_+$,

$$\sup_E Gf = \sup_{\{f>0\}} Gf. \tag{3.16}$$

(ii) If $A \subseteq B$, then

$$G'_A = I_{B^c} G'_B + G'_A I_B G'_B, \tag{3.17}$$

$$U_A = U_B + U_A I_{B \setminus A} U_B. \tag{3.18}$$

Proof. (i) In order to prove (3.13) use (3.11) and the strong Markov property at T_B. Formula (3.14) then follows by 'multiplying' both sides of (3.13) from the left by P, and using (3.8) and (3.9). Formula (3.15) is obtained by setting $A = \varnothing$ in (3.13). To get (3.16) use (3.15) with $B = \{f > 0\}$.

(ii) In order to prove (3.17), note first that, for $A \subseteq B$, and $n \geq 0$:

$$\{S_A > n\} = \{T_B > n\} + \sum_{m=0}^{n} \{S_A > m, L_B(n) = m\},$$

where

$$L_B(n) = \max\{m: 0 \leq m \leq n, X_m \in B\}.$$

Consequently, for any $f \in \mathscr{E}_+$,

$$\begin{aligned} G'_A f(x) &= \sum_{n=0}^{\infty} \mathbb{E}_x[f(X_n); S_A > n] \\ &= \sum_{n=0}^{\infty} \mathbb{E}_x[f(X_n); T_B > n] \\ &\quad + \sum_{m=0}^{\infty} \sum_{n=m}^{\infty} \mathbb{E}_x[f(X_n); S_A > m, L_B(n) = m] \end{aligned}$$

The first term on the right hand side equals $I_{B^c} G'_B f(x)$. The second is equal to

$$\sum_{m=0}^{\infty} \mathbb{E}_x\left[1_B(X_m) \mathbb{E}_{X_m}\left[\sum_{n=0}^{S_B - 1} f(X_n) \right]; S_A > m \right] = G'_A I_B G'_B f(x),$$

which is just what we wanted.

Formula (3.18) follows after 'multiplying' (3.17) from the right by P. □

The equations (3.14) and (3.18) are called the *resolvent equations* while (3.16) is usually referred to as the *maximum principle.*

We denote by L_B the *last exit time* from the set B:

$$L_B = \sup\{n \geq 0: X_n \in B\}.$$

Of course L_B is defined only on the set $\{T_B < \infty\} = \{X_n \in B \text{ for some } n \geq 0\}$. Obviously

$$\{L_B = n\} = \{X_n \in B,\ \ S_B \circ \theta_n = \infty\},\ \ n \geq 0,$$
$$\{L_B = \infty\} = \{X_n \in B \quad \text{i.o.}\}.$$

Note the special case

$$L_E = \sup\{n \geq 0: X_n \in E\} = \inf\{n \geq 0: X_{n+1} = \Delta\}.$$

L_E is called the *lifetime* of (X_n).

Let us define

$$h_B(x) = \mathbb{P}_x\{T_B < \infty\} = G_B(x, B) \qquad \text{(cf. 3.11)},$$
$$p_B(x) = \mathbb{P}_x\{0 \leq L_B < \infty\},$$
$$g_B(x) = \mathbb{P}_x\{L_B = 0\} = 1_B(x)\mathbb{P}_x\{S_B = \infty\}$$
$$= I_B(1 - U_B 1_B)(x) \qquad \text{(cf. 3.12)},$$

$$h_B^\infty(x) = \mathbb{P}_x\{X_n \in B \quad \text{i.o.}\} = \lim_{n\to\infty} \downarrow (U_B I_B)^n 1(x). \tag{3.19}$$

Theorem 3.4. (i) The function h_B is superharmonic, p_B is the potential with charge g_B, h_B^∞ is harmonic, and (p_B, h_B^∞) is the Riesz decomposition of h_B. Moreover, h_B is the smallest superharmonic function h satisfying $h \geq 1_B$.

(ii) The function h_B^∞ is also harmonic for the transition probability $U_B I_B(x, \mathrm{d}y) = \mathbb{P}_x\{X_{S_B} \in \mathrm{d}y,\ S_B < \infty\}$. It is the harmonic part in the Riesz decomposition of the superharmonic function 1 (for the transition probability $U_B I_B$).

Proof. (i) By the definitions of L_B and g_B, and by the Markov property,

$$\mathbb{P}_x\{L_B = n\} = \mathbb{P}_x\{L_B \circ \theta_n = 0\} = P^n g_B(x) \quad \text{for all } n \geq 0.$$

Summing over n gives

$$p_B = G g_B,$$

i.e. p_B is the potential with charge g_B.

For the function h_B we get

$$h_B(x) = \mathbb{P}_x\{T_B < \infty\} = \mathbb{P}_x\{L_B = 0\} + \mathbb{P}_x\{S_B < \infty\}$$
$$= g_B(x) + P h_B(x),$$

and

$$\lim_{n\to\infty} \downarrow P^n h_B(x) = \lim_{n\to\infty} \downarrow \mathbb{P}_x\{T_B \circ \theta_n < \infty\}$$
$$= \mathbb{P}_x\{X_n \in B \quad \text{i.o.}\} = h_B^\infty(x).$$

This proves that (p_B, h_B^∞) is the Riesz decomposition of the superharmonic function h_B.

In order to prove the minimality of h_B, apply Theorem 3.1(iii) with $g = 1_B$ and $K = I_{B^c}P$.

(ii) The result follows directly from the representation $h_B^\infty = \lim_{n\to\infty}\downarrow (U_B I_B)^n 1$, see (3.19). □

We write

$$B^\infty = \{h_B^\infty = 1\} = \{x \in E : \mathbb{P}_x\{X_n \in B \quad \text{i.o.}\} = 1\}.$$

From the preceding theorem and Propositions 3.2 and 3.3 we obtain:

Proposition 3.8. For any sets $A, B \in \mathscr{E}$:

(i) Either $h_B^\infty > 0$ everywhere or the set $\{h_B^\infty = 0\}$ is closed.

(ii) Either $h_B^\infty < 1$ everywhere or the set $B^\infty = \{h_B^\infty = 1\}$ is absorbing. If $B^\infty \neq \varnothing$, then $B \cap B^\infty \neq \varnothing$ and $B \cap B^\infty \subseteq (B \cap B^\infty)^\infty$.

(iii) Either $h_B^\infty \equiv 0$, or $\sup_B h_B^\infty = 1$.

(iv) If $B^\infty \neq \varnothing$ and $\inf_B h_A > 0$ then $B^\infty \subseteq A^\infty \neq \varnothing$.

(v) If B is absorbing then $p_B \equiv 0$ and $h_B = h_B^\infty$.

Proof. (i) Use Proposition 3.2(ii).

(ii) If $B^\infty \neq \varnothing$ then it is absorbing by Proposition 3.2(iv). Clearly, if $B^\infty \neq \varnothing$ then it is impossible that $B \cap B^\infty = \varnothing$. If now $X_0 = x \in B \cap B^\infty$, then $X_n \in B^\infty$ for all $n \geq 0$ and $X_n \in B$ infinitely often, whence $X_n \in B \cap B^\infty$ infinitely often (a.s.). Thus $B \cap B^\infty \subseteq (B \cap B^\infty)^\infty$ as asserted.

(iii) If $h_B^\infty \leq \rho < 1$ on B, then by Theorem 3.4(ii), $h_B^\infty = U_B I_B h_B^\infty \leq \rho$ everywhere, which by Proposition 3.3 leads to $h_B^\infty \equiv 0$.

(iv) Define a transition probability Q by

$$\begin{aligned} Q(x, \mathrm{d}y) &= \mathbb{P}_x\{X_{S_B} \in \mathrm{d}y, S_B < \infty, S_B \leq T_A \leq \infty\} \\ &= \mathbb{P}_x\{X_0 \notin A, X_{S_{A\cup B}} \in B \cap \mathrm{d}y, S_{A\cup B} < \infty\} \\ &= I_{A^c} U_{A\cup B} I_B(x, \mathrm{d}y). \end{aligned}$$

Clearly the set B^∞ is closed for Q. For any $x \in B^\infty$, $i \geq 1$, we have $\mathbb{P}_x\{S_B(i) < \infty\} = 1$ and $Q^i 1(x) = \mathbb{P}_x\{T_A \geq S_B(i)\}$, and hence

$$\lim_{i\to\infty}\downarrow Q^i 1(x) = \mathbb{P}_x\{T_A = \infty\} = 1 - h_A(x).$$

It follows that the function $1 - h_A$ is the harmonic part in the Riesz decomposition of the superharmonic function 1 (for the transition probability Q restricted to the closed set B^∞). Let $\rho = 1 - \inf_B h_A < 1$. On B^∞ we have

$$1 - h_A = Q(1 - h_A) = I_{A^c} U_{A \cup B} I_B (1 - h_A) \leq \rho.$$

Consequently, by Proposition 3.3, $1 - h_A = 0$ on B^∞. By Theorem 3.4(i), and since B^∞ is absorbing,

$$h_A^\infty = \lim_{n \to \infty} P^n h_A \geq \lim_{n \to \infty} P^n 1_{B^\infty} \geq 1_{B^\infty},$$

from which it follows that $B^\infty \subseteq A^\infty$.

(v) Obvious. □

3.5 The dissipative and conservative parts

Our goal in this section is to obtain the decomposition $E = E_c + E_d$ mentioned at the beginning of this chapter.

Definition 3.3. A set $B \in \mathscr{E}$ is called *transient*, if

$$\mathbb{P}_x\{L_B < \infty\} = 1 \quad \text{for all } x \in B,$$

(or, equivalently, $h_B^\infty = 0$ on B).

B is called *dissipative*, if there exists a function $g \in \mathscr{E}_+$ such that

$$g > 0 \text{ and } Gg < \infty \text{ on } B.$$

If the whole state space E is transient (resp. dissipative), we say that the Markov chain (X_n) is transient (resp. dissipative).

Note that, by Proposition 3.8(iii), if B is transient then in fact $h_B^\infty \equiv 0$.

Note also that (X_n) is transient if and only if the lifetime L_E of (X_n) is finite $\mathbb{P}_x$-a.s. for all $x \in E$.

The relationships between the concepts of transience and dissipativity are discussed in the following:

Proposition 3.9. Let $B \in \mathscr{E}$ be arbitrary.

(i) If $G1_B$ is finite on B then B is transient.

(ii) The sets $\{p_B > 0\} = \{h_B^\infty < h_B\}$ and $B \backslash B^\infty = \{x \in B : h_B^\infty(x) < 1\}$ are dissipative. In particular, if B is transient, then it is dissipative.

(iii) If B is dissipative, then there exists $g \in \mathscr{E}_+$ such that $\{g > 0\} = B$ and $\sup_E Gg \leq 1$.

(iv) If B is dissipative then $B = \bigcup_{i=1}^\infty B_i$ where each B_i is transient.

(v) If $B_i, i \geq 1$, are dissipative then so is their union $B = \bigcup_{i=1}^\infty B_i$.

(vi) If B is dissipative then there exists a dissipative set $\bar{B} \supseteq B$ such that the complement $(\bar{B})^c$ is either empty or absorbing.

(vii) If (X_n) is irreducible dissipative, then the potential Gs is bounded for any small $s \in \mathscr{S}^+$. Then also every small set C is transient.

Proof. (i) Since (by hypothesis) $\mathbb{E}_x \sum_0^\infty 1_B(X_n) < \infty$, we have

$$h_B^\infty(x) = \mathbb{P}_x\left\{\sum_0^\infty 1_B(X_n) = \infty\right\} = 0 \quad \text{for all } x \in B.$$

(ii) Set

$$g = \sum_0^\infty 2^{-(n+1)} P^n g_B.$$

By Theorem 3.4(i)

$$\sum_0^\infty P^n g_B = G g_B = p_B = h_B - h_B^\infty,$$

whence $g > 0$ on $\{h_B^\infty < h_B\}$. Moreover

$$Gg = \sum_0^\infty 2^{-(n+1)} G P^n g_B \leq \sum_0^\infty 2^{-(n+1)} G g_B \leq 1.$$

Thus $\{h_B^\infty < h_B\}$ and hence also its subset $B \backslash B^\infty$ are dissipative.

(iii) and (iv): Let $g \in \mathscr{E}_+$ be such that $\{g > 0, Gg < \infty\} = B$, and let

$$B_i = \{g \geq i^{-1}, Gg \leq i\}, \quad i \geq 1.$$

Then $B = \bigcup_1^\infty B_i$ and by the maximum principle (see (3.16))

$$\sup_E G1_{B_i} = \sup_{B_i} G1_{B_i} \leq i^2.$$

By (i) each B_i is transient. The desired function g in (iii) is given by

$$g = 2^{-i} i^{-2} \text{ on } B_i \backslash \bigcup_1^{i-1} B_j, \quad g = 0 \text{ on } B^c.$$

(v) Let $g_i \in \mathscr{E}_+$ be such that $\{g_i > 0\} = B_i$ and $Gg_i \leq 1$. Set $g = \sum_1^\infty 2^{-i} g_i$. Then we have $\{g > 0\} = B$ and $Gg \leq 1$.

(vi) First we show that the set $B^+ = \{h_B > 0\} \supseteq B$ is also dissipative. To this end, let $g \in \mathscr{E}_+$ be such that $\{g > 0\} = B$ and $Gg \leq 1$, and define

$$f = \sum_0^\infty 2^{-(n+1)} P^n g.$$

Then $\{f > 0\} = B^+$ and $Gf \leq 1$, which shows that B^+ is dissipative.

Now recall from (ii) that the set $B^c \backslash (B^c)^\infty$ is dissipative. It follows that the union

$$\bar{B} = B^+ \cup (B^c \backslash (B^c)^\infty)$$

is also dissipative. Its complement is equal to

$$(\bar{B})^c = B^0 \cap (B^c)^\infty.$$

If $(\bar{B})^c$ is not empty, then, being the intersection of a closed and of an absorbing set (see Propositions 2.2 and 3.8 (ii)), it is absorbing.

(vii) Use (iii) and Proposition 2.7(ii). The latter result then follows from (i). □

From Theorem 3.1(iii) we obtain the following criteria for dissipativity and transience:

Proposition 3.10. Let $B \in \mathscr{E}$ be arbitrary.

(i) B is dissipative if and only if there exists a superharmonic function h satisfying

$$\infty > h > Ph \text{ on } B.$$

(ii) If there exists a superharmonic function h and a constant $\gamma > 0$ such that

$$\infty > h \geq Ph + \gamma \text{ on } B,$$

then B is transient.

Proof. Let $g = h - Ph$ and apply Theorem 3.1(iii) and Proposition 3.9(i). □

Example 3.2. (*a*) Suppose that $(X_n; n \geq 0)$ is a discrete Markov chain (cf. Example 1.2(*a*)). Call a state $x \in E$ transient (resp. dissipative), if $\{x\}$ is a transient (resp. dissipative) set. Now any subset $B \subseteq E$ is dissipative if and only if every state $x \in B$ is transient. In particular, an individual state $x \in B$ is transient if and only if it is dissipative.

Let φ be an arbitrary non-trivial σ-finite measure on $(E, \mathscr{E})$.

Definition 3.4. A φ-positive set $B \in \mathscr{E}$ is called *φ-conservative*, if for all φ-positive sets $A \subseteq B$,

$$h_A^\infty = 1 \quad \varphi\text{-a.e. on } A. \tag{3.20}$$

If the whole state space E is φ-conservative we say that the Markov chain (X_n) is φ-conservative.

Note that, if $\varphi P \ll \varphi$ then by (3.19) the condition

$$\mathbb{P}_x\{S_A < \infty\} = U_A(x, A) = 1 \quad \text{for } \varphi\text{-a.e. } x \in A$$

is sufficient for (3.20).

φ-conservative sets can be characterized as being complementary to all φ-positive dissipative sets:

Proposition 3.11. A φ-positive set $B \in \mathscr{E}$ is φ-conservative if and only if it does not contain any φ-positive dissipative set $A \in \mathscr{E}$.

Proof. If a φ-positive set $A \subseteq B$ is dissipative, then, by Proposition 3.9 (iv), it contains a φ-positive transient set. Hence B is not φ-conservative.

The converse result is a direct consequence of Proposition 3.9 (ii). □

After all these preliminaries the proof of the decomposition result will be easy.

Theorem 3.5. (Hopf's decomposition). Let $\varphi \in \mathscr{M}^+$ be a σ-finite measure. The state space $(E, \mathscr{E})$ can be divided into two parts

$$E = E_d(\varphi) + E_c(\varphi),$$

where $E_d(\varphi)$ is dissipative, and we have either

(i) $E_c(\varphi) = \varnothing$ (i.e., the Markov chain (X_n) is dissipative), or
(ii) $\varphi(E_c(\varphi)) = 0$, and $E_c(\varphi)$ is absorbing, or
(iii) $\varphi(E_c(\varphi)) > 0$, and $E_c(\varphi)$ is absorbing and φ-conservative.

Proof. Let $B_i, i \geq 1$, be dissipative sets such that their union $B = \bigcup_1^\infty B_i$, which by Proposition 3.9(v) is also dissipative, is a version of φ-ess sup $\{A \in \mathscr{E} : A \text{ dissipative}\}$. Let $\bar{B} \supseteq B$ be defined as in Proposition 3.9(vi), and set $E_d(\varphi) = \bar{B}$, $E_c(\varphi) = (\bar{B})^c$. The result now follows from Propositions 3.9(vi) and 3.11. □

Examples 3.3. (*a*) Suppose that (X_n) is a discrete Markov chain and let $\varphi = \text{Card}$ be the counting measure on E. Call a state $x \in E$ recurrent, if it is Card-conservative; this means

$$h_x^\infty(x) = \mathbb{P}_x\{X_n = x \quad \text{i.o.}\} = 1,$$

or equivalently, see Example 3.2(*a*),

$$g(x,x) = \sum_{n=0}^{\infty} p^n(x,x) = \infty.$$

Also, x is recurrent if and only if it is not transient. Consequently, $E_d = E_d(\text{Card}) = \{x \in E\colon x \text{ transient}\}$ and $E_c = E_c(\text{Card}) = \{x \in E\colon x \text{ recurrent}\}$.

(*b*) (The forward process). Suppose that F is non-lattice. Let $\bar{M} = \text{ess sup } z_1 = \sup\{t : F(t) < 1\}$. If $M = \infty$, then $(V_{n\delta}^+; n \geq 0)$ is ℓ-conservative. If $\bar{M} < \infty$, then $E_d(\ell) = (\bar{M}, \infty)$ and $E_c(\ell) = [0, \bar{M}]$.

3.6 Recurrence

Recall from Chapter 2 the definitions of indecomposability and irreducibility. In particular, recall that the latter implies the former. It is easy to see that the converse is not true in general. (Take, for example, the discrete Markov chain on the integers moving deterministically one step to the right.)

We shall now examine what form Theorem 3.5 takes when we assume that the state space E is indecomposable. It turns out that there is a strong dichotomy: the Markov chain (X_n) is either dissipative or irreducible 1-recurrent.

When dealing with irreducible kernels and Markov chains we adopt the notation and terminology used in Chapter 2. In particular, ψ denotes a maximal irreducibility measure, and $\mathscr{E}^+ = \{B \in \mathscr{E} : \psi(B) > 0\}$.

We introduce the following:

Definition 3.5. An irreducible Markov chain (X_n) is called *recurrent*, if

$$h_B^\infty > 0 \quad \text{everywhere and}$$
$$h_B^\infty = 1 \ \psi\text{-almost everywhere, for all } B \in \mathscr{E}^+,$$

and *Harris recurrent* (or ψ-recurrent), if

$$h_B^\infty \equiv 1 \quad \text{for all } B \in \mathscr{E}^+.$$

Note that, if (X_n) is recurrent then it is ψ-conservative. Note also that then the set $\{h_B^\infty = 1\}$ is absorbing (see Proposition 3.8(ii)) for all $B \in \mathscr{E}^+$; in particular, if (X_n) is Harris recurrent then P is stochastic. As one might guess, recurrence must be closely related to the concept of 1-recurrence. In fact, we will soon see that they actually coincide.

For the following theorem, recall from (2.1) that there always exist measures φ satisfying $\varphi P \ll \varphi$ (see also Lemma 2.2).

Theorem 3.6. (i) Suppose that the state space E is indecomposable. Then the Markov chain (X_n) is either dissipative or (irreducible) recurrent.

(ii) Suppose that (X_n) is recurrent. Then, for any non-trivial σ-finite measure φ on $(E, \mathscr{E})$ satisfying $\varphi P \ll \varphi$, the restriction of φ to the conservative part $E_c(\varphi)$ is a maximal irreducibility measure; i.e.,

$$\varphi I_{E_c(\varphi)} \sim \psi.$$

Proof. Suppose that (X_n) is not dissipative. Let φ be such that $\varphi P \ll \varphi$. By Lemma 2.2(ii) we necessarily are in case (iii) of Theorem 3.5. Let $B \subseteq E_c(\varphi)$ be an arbitrary φ-positive set. By Proposition 3.8(ii) the set $B^\infty = \{h_B^\infty = 1\}$ is absorbing. (B^∞ is non-empty since $E_c(\varphi)$ is conservative.) By the same proposition (part (i)) and by our assumption of indecomposability, $h_B^\infty > 0$ everywhere. In particular, we see that (X_n) is $\varphi I_{E_c(\varphi)}$-irreducible. It follows from Proposition 2.4(ii) that the irreducibility measure $\varphi I_{E_c(\varphi)}$ is maximal. □

In the following theorem we give several characterizations for recurrence.

Theorem 3.7. Suppose that the Markov chain (X_n) is irreducible.

(X_n) is recurrent if and only if any of the following equivalent conditions (i)–(vi) is satisfied:

(i) there exists an absorbing full set H such that (X_n) restricted to H is Harris recurrent;

(ii) the transition probability P is 1-recurrent;

(iii) for some small function $s \in \mathscr{S}^+$, the potential Gs is unbounded;

(iv) (X_n) is ψ-conservative;

(v) for some small set $C \in \mathscr{S}^+$, the set C^∞ is non-empty;

(vi) for some small set $C \in \mathscr{S}^+, U_C 1_C = 1 \quad \psi$-a.e. on C.

(X_n) is Harris recurrent if (and only if):

(vii) for some small set $C \in \mathscr{S}^+, U_C 1_C \equiv 1$.

Proof. The implications (i)$\Rightarrow$recurrence$\Rightarrow$(ii)$\Rightarrow$(iii) and recurrence$\Rightarrow$(iv)$\Rightarrow$(v)$\Rightarrow$(iii) are obvious. That (iii) implies recurrence follows from Proposition 3.9(vii) and part (i) of the preceding theorem. Condition (v) implies (vi), since $C^\infty \neq \varnothing$ is absorbing and full; the converse was noted already in the remark after Definition 3.4.

In order to prove the remaining parts of the theorem, let us fix a small set $C \in \mathscr{S}^+$ and write $H = C^\infty$. In any case it follows from Propositions 2.7(i) and 3.8(iv) that $H \subseteq B^\infty$ for all $B \in \mathscr{E}^+$. This proves that (v) implies (i), and (vii) implies the Harris recurrence (note that $h_C^\infty = \lim_{n\to\infty} \downarrow (U_C I_C)^n 1 \equiv 1$ when (vii) holds). So the whole proof is completed. □

We call an absorbing set $H \in \mathscr{E}$ such that (X_n) restricted to H is Harris recurrent, a *Harris set for* (X_n).

From Theorems 3.2 and 3.7 and Proposition 3.9(vii) we immediately obtain the following:

Corollary 3.1. For an irreducible Markov chain (X_n):

(i) (X_n) is dissipative if and only if $R > 1$ or P is 1-transient.

(ii) If $R > 1$ or P is 1-transient, the potentials Gs, $s \in \mathscr{S}^+$, are bounded. □

According to the following proposition, the (seemingly weaker) notion of φ-recurrence is equivalent to Harris recurrence:

Proposition 3.12. Let $\varphi \in \mathscr{M}^+$ be an arbitrary σ-finite measure. If (X_n) is φ-*recurrent*, that means

$$h_B^\infty \equiv 1 \quad \text{for all } \varphi\text{-positive } B \in \mathscr{E},$$

then (X_n) is Harris recurrent (and $\varphi \ll \psi$).

Proof. Clearly, (X_n) is φ-irreducible (whence $\varphi \ll \psi$). The assertion now follows by taking a φ-positive small set $C \in \mathscr{S}^+$ and using criterion (vii) of Theorem 3.7. □

It turns out that Harris recurrence is closely related to the following condition: Every bounded harmonic function is a constant.

Theorem 3.8. (i) If the Markov chain (X_n) is Harris recurrent then for any superharmonic function h there is a constant $0 \le c < \infty$ such that $h = c$ ψ-a.e. and $h \ge c$ everywhere. Then also every bounded harmonic function is a constant.

(ii) Conversely, if every bounded harmonic function is a constant, then (X_n) is either dissipative or Harris recurrent.

Proof. (i) Suppose that $h \in \mathscr{E}_+$ is superharmonic. Let $c = \psi$-ess sup h. Then for every $\varepsilon > 0$, the set $B = \{h \ge c - \varepsilon\}$ is ψ-positive. From Theorem 3.4(i) and from our hypothesis it follows that $h \ge (c - \varepsilon) h_B \equiv c - \varepsilon$. Thus $h \ge c$ everywhere.

Suppose now that $h \in b\mathscr{E}_+$ is harmonic. Applying the result we just proved to the harmonic functions h and $\sup_E h - h$ we see that h is a constant.

(ii) The result follows directly from the fact that h_B^∞ is harmonic for all $B \in \mathscr{E}$, and from Theorem 3.6(i). □

For recurrent chains we have:

Proposition 3.13. Suppose that (X_n) is recurrent. Then:

(i) Every superharmonic function is constant ψ-a.e.

(ii) There exists a harmonic function $\underline{h}$, $\underline{h} = 1$ ψ-a.e., $\underline{h} \le 1$, which is minimal among the superharmonic functions h satisfying $h = 1$ ψ-a.e.; $\underline{h}$ is given by

$$\underline{h} = h_H = h_H^\infty \quad \text{for all Harris sets } H \in \mathscr{E}.$$

(iii) There exists a Harris set $\bar{H} \in \mathscr{E}$ which is maximal in the sense that $H \subseteq \bar{H}$ for every Harris set $H \in \mathscr{E}$. Also, if $h \in \mathscr{E}_+$, $h = 1$ ψ-a.e., is superharmonic, then $h \ge 1$ on $\bar{H}$. $\bar{H}$ is given by

$$\bar{H} = \{\underline{h} = 1\} = \{h_H = 1\} = H^\infty = C^\infty$$

for all Harris sets $H \in \mathscr{E}$, all small sets $C \in \mathscr{S}^+$.

Proof. (i) This follows directly from Theorems 3.7(i) and 3.8(i).

(ii) Let $H \in \mathscr{E}$ be a Harris set. By Proposition 3.8(v) $h_H = h_H^\infty$ is harmonic. By Theorem 3.8(i), if h is any superharmonic function satisfying $h = 1$ ψ-a.e., then $h \ge 1$ on H. By Theorem 3.4(i) this implies that $h \ge h_H^\infty$.

(iii) This is a direct consequence of (ii), and of the fact that C^∞ is a Harris set satisfying $(C^\infty)^\infty = C^\infty$ (cf. the proof of Theorem 3.7). □

Next we shall look at the recurrence of the m-step Markov chains

$(X_{nm}; n \geq 0)$, $m \geq 1$. As a direct corollary of Proposition 3.5 and Corollary 3.1 we have:

Corollary 3.2. Suppose that (X_n) is irreducible. Then (X_n) is recurrent if and only if some (or equivalently, all) of the $c_m = \text{g.c.d.}\{m, d\}$ m-step chains $(X_{nm}; n \geq 0)$ is (are) recurrent. □

Concerning the Harris recurrence of the m-step chains we obtain the following result:

Proposition 3.14. Suppose that (X_n) is Harris recurrent. Then for any $m \geq 1$, all the c_m m-step chains $(X_{nm}; n \geq 0)$ are Harris recurrent.

Proof. Since (X_n) restricted to the closed set $E_0 + \cdots + E_{d-1} = N^c$ is Harris recurrent there is no loss of generality in supposing that N is empty. Let $\underline{h}_i$, $\underline{h}_i = 1$ ψ-a.e. on $E_i^{(m)}$, $\underline{h}_i = 0$ on $(E_i^{(m)})^c$, be the minimal harmonic function (given by Proposition 3.13(ii)) for the Markov chain (X_{nm}) with state space $E_i^{(m)}$. Thus

$$P^m \underline{h}_i = \underline{h}_i, \quad i = 0, 1, \ldots, c_m - 1. \tag{3.21}$$

It follows that $P\underline{h}_i$ is harmonic for (X_{nm}) with state space $E_{i-1}^{(m)}$. By minimality

$$P\underline{h}_i \geq \underline{h}_{i-1}.$$

(The indices are modulo c_m.) By iterating we get

$$P^{c_m}\underline{h}_i \geq P^{c_m - 1}\underline{h}_{i+c_m-1} \geq \cdots \geq P\underline{h}_{i+1} \geq \underline{h}_i.$$

Further iterating leads to

$$P^m \underline{h}_i \geq \cdots \geq P^{2c_m}\underline{h}_i \geq P^{c_m} h_i \geq h_i.$$

It follows from (3.21) that all the above inequalities are equalities. In particular,

$$P\underline{h}_i = \underline{h}_{i-1} \quad \text{for all } i = 0, 1, \ldots, c_m - 1,$$

from which we can conclude that the function h defined by

$$h = \sum_0^{c_m - 1} \underline{h}_i \; (= \underline{h}_i \text{ on } E_i^{(m)}),$$

is harmonic. By Theorem 3.8(i), $h \equiv 1$, and hence also $\underline{h}_i = 1$ on $E_i^{(m)}$ for all i. The final result now follows from Proposition 3.13(iii). □

Examples 3.4. (*a*) Let (X_n) be a discrete, irreducible Markov chain (cf. Example 2.2(*a*)). (X_n) is recurrent if and only if there exists a recurrent state $x_0 \in E$. Then all the states x in $F = \{x \in E : x_0 \to x\}$ are recurrent, and F is a

Harris set for (X_n). The maximal Harris set $\bar{H}$ is given by

$$\bar{H} = \{x \in E : h_{x_0}(x) \stackrel{\text{def}}{=} \mathbb{P}_x\{X_n = x_0 \quad \text{for some } n \geq 0\} = 1\}.$$

In particular, (X_n) is Harris recurrent if and only if $h_{x_0} \equiv 1$ for some recurrent state x_0.

(*c*) (The random walk). Suppose that F is non-lattice, and has mean

$$M_F = \int_{\mathbb{R}} tF(\mathrm{d}t).$$

If $M_F \neq 0$, then the random walk (Z_n) visits every finite interval only finitely many times almost surely. On the other hand, if $M_F = 0$, then every interval of positive length is visited infinitely often almost surely (see e.g. Feller (1971), Sect. VI. 10). It follows that the random walk is dissipative if $M_F \neq 0$, and Harris recurrent if F is spread-out and $M_F = 0$.

(*d*) (The reflected random walk). Suppose that F has a mean M_F, $-\infty \leq M_F \leq \infty$. The reflected random walk (W_n) is dissipative if and only if $M_F > 0$, and Harris recurrent if and only if $M_F \leq 0$.

(*e*) (The forward process). Suppose that F is spread-out. Then, for any $\delta > 0$, the Markov chain $(V^+_{n\delta})$ is Harris recurrent.

(*f*) With the assumptions of Example 2.1(*f*), the autoregressive process (R_n) is Harris recurrent.

4

Embedded renewal processes

In this chapter we shall develop an important tool for the study of irreducible kernels, the so-called regeneration method.

It can be most easily described in the case where $K = P$ is the transition probability of a Markov chain (X_n) having a *communicating state*; that is a state $x_0 \in E$ such that $\{x_0\} \in \mathscr{E}$ and $x_0 \to x_0$, i.e.,

$$P^n(x_0, \{x_0\}) > 0 \quad \text{for some } n \geq 1.$$

Then the path of the chain splits in a natural manner into x_0-*blocks*; these are the subpaths between consecutive visits to x_0 by (X_n). By the Markov property the x_0-blocks are i.i.d. random elements, and thus a single x_0-block contains all the relevant 'information' about the whole path of the chain. Since the chain can be considered to regenerate at every visit to x_0, this method is commonly referred to as the *regeneration method.*

We shall show in this chapter that the regeneration method has far wider applicability than would be expected at first sight. In fact, we will see that a sufficient condition for the successful application of this method is that we have an irreducible kernel K which satisfies the minorization condition $M(m_0, \beta, s, \nu)$ with $m_0 = 1$, i.e.,

$$K \geq \beta s \otimes \nu.$$

We usually normalize β and s so that $\beta = 1$, and then call the pair (s, ν) an *atom* for K.

By Theorem 2.1, for an irreducible kernel K, some iterate $K^{m_0}, m_0 \geq 1$, possesses an atom (s, ν). Therefore the regeneration method is applicable for the entire class of irreducible kernels.

4.1 Renewal sequences and renewal processes

Since renewal theory will play a central role in the sequel we start by proving some elementary results for renewal sequences and processes.

Let $b = (b_n; n \geq 0)$ be a sequence satisfying

$$b_0 = 0, \quad 0 \leq b_n < \infty \quad \text{for all } n \geq 0, \quad b_n > 0 \quad \text{for some } n \geq 1. \quad (4.1)$$

Let $b^{*i}, i \geq 0$, denote the convolution powers of b, i.e., we set

$$b^{*0} = \delta = (\delta_n; n \geq 0) \overset{\text{def}}{=} (1, 0, 0, \ldots),$$

and iteratively,

$$b^{*i} = b * b^{*(i-1)} \quad \text{for } i \geq 1.$$

(The convolution $a * a' = (a * a'_n; n \geq 0)$ of any two sequences $a = (a_n; n \geq 0)$ and $a' = (a'_n; n \geq 0)$ is defined in the usual manner:

$$a * a'_n = \sum_{m=0}^{n} a_m a'_{n-m}, \quad n \geq 0.)$$

Note that from our hypothesis $b_0 = 0$ it follows that for every $i \geq 1$,

$$b_n^{*i} = 0 \quad \text{for } 0 \leq n < i. \tag{4.2}$$

Definition 4.1. The sequence $u = (u_n; n \geq 0)$ defined by

$$u = \sum_{i=0}^{\infty} b^{*i}$$

is called the (undelayed) *renewal sequence corresponding to the increment sequence b.*

If $a = (a_n; n \geq 0)$ is an arbitrary non-negative sequence, the sequence

$$v = a * u = a * \sum_{0}^{\infty} b^{*i}$$

is called a *delayed renewal sequence* (with delay sequence a).

Here are some elementary facts about renewal sequences:

Proposition 4.1. (i) $u_0 = 1$, $0 \leq u_n < \infty$ for all $n \geq 0$.

(ii) g.c.d. $\{n \geq 1: b_n > 0\}$ = g.c.d. $\{n \geq 1: u_n > 0\}$.

(iii) The renewal sequence u is the unique non-negative sequence $u = (u_n; n \geq 0)$ satisfying the equation

$$u = \delta + b * u. \tag{4.3}$$

More generally, the delayed renewal sequence $v = a * u$ is the unique non-negative sequence $v = (v_n; n \geq 0)$ satisfying the equation

$$v = a + b * v. \tag{4.4}$$

Proof. (i) By (4.2), we have $u_n = \sum_{i=0}^{n} b_n^{*i}$.

(ii) The result is obvious.

(iii) Let a be an arbitrary non-negative sequence and let $v = a * u$. Clearly,

$$v = a + a * \sum_{1}^{\infty} b^{*i} = a + b * v.$$

If v' is any non-negative solution of (4.4), then we obtain by iterating

$$v'_n = a_n + v' * b_n = \sum_{i=0}^{j} a * b_n^{*i} + v' * b_n^{*(j+1)} \quad \text{for all } n, j \geq 0.$$

If $j \geq n$ then by (4.2) the first term on the right hand side is equal to $a*u_n$ and the second is zero. □

The integer $d \geq 1$ defined by

$$d = \text{g.c.d.}\{n \geq 1: b_n > 0\} = \text{g.c.d.}\{n \geq 1: u_n > 0\}$$

is called the *period* of the renewal sequence u. In most cases we do not lose the generality if we suppose that u is *aperiodic*, that is, $d = 1$. (Otherwise we should consider the aperiodic renewal sequence $(u_{nd}\,; n \geq 0)$ corresponding to the increment sequence $(b_{nd}\,; n \geq 0)$.)

Equation (4.3) is called the *renewal equation*.

If $M_b^{(0)} \overset{\text{def}}{=} \sum_1^\infty b_n \leq 1$, we call the corresponding renewal sequence u *probabilistic*. In this case we have the following interpretation:

Let $t(i)$, $1 \leq t(i) \leq \infty$ a.s., $i = 1, 2, \ldots$, be i.i.d. random times with common distribution b. Write briefly $t(1) = t$. Thus we have, for any $i \geq 1$,

$$\mathbb{P}\{t(i) = n\} = \mathbb{P}\{t = n\} = b_n, \quad n \geq 1,$$

$$\mathbb{P}\{t(i) = \infty\} = \mathbb{P}\{t = \infty\} = 1 - M_b^{(0)}.$$

Let $T(0)$, $0 \leq T(0) \leq \infty$ a.s., be a random time, independent of $\{t(i); i \geq 1\}$. We call $T(0)$ the *delay*. Let $a = (a_n\,; n \geq 0)$ be its distribution:

$$\mathbb{P}\{T(0) = n\} = a_n, \quad n \geq 0,$$

$$\mathbb{P}\{T(0) = \infty\} = 1 - \sum_0^\infty a_n = 1 - M_a^{(0)}.$$

We write $T(i), i \geq 1$, for the sums

$$T(i) = T(0) + \sum_{j=1}^{i} t(j).$$

Definition 4.2. The sequence $(T(i); i \geq 0)$ defined above is called the *renewal process* (on $\mathbb{N}$) *with delay distribution a and increment distribution b*. The sequence $(Y_n\,; n \geq 0)$ defined by

$$Y_n = 1_{\{T(i) = n \text{ for some } i \geq 0\}}$$

is called the *incidence process* (of the renewal process $(T(i); i \geq 0)$).

If $a = \delta$, i.e., $T(0) = 0$ a.s., then the renewal process $(T(i); i \geq 0)$ is called *undelayed*.

In order to emphasize a specific delay distribution a we will often write $\mathbb{P}_a$ instead of $\mathbb{P}$ for the underlying probability measure.

It is easy to see that

$$v_n = a*u_n = \mathbb{P}_a\{Y_n = 1\}.$$

In particular, for the undelayed renewal sequence u we have

$$u_n = \mathbb{P}_\delta\{Y_n = 1\}.$$

Note that in the probabilistic case

$$0 \le u_n \le 1 \quad \text{for all } n \ge 0,$$

and

$$\sum_0^\infty u_n = \sum_{n=0}^\infty \sum_{i=0}^\infty b_n^{*i} = \sum_{i=0}^\infty (M_b^{(0)})^i = (1 - M_b^{(0)})^{-1}. \tag{4.5}$$

An undelayed renewal process $(T(i); i \ge 0)$ (or the corresponding renewal sequence u) is called *recurrent* if $T(i) < \infty$ a.s. for all $i \ge 0$, otherwise *transient.*

Proposition 4.2. Either of the following conditions is equivalent to the recurrence of the renewal process $(T(i); i \ge 0)$:
(i) $M_b^{(0)} = 1$,
(ii) $\sum_0^\infty u_n = \infty$.

Proof. Obviously, recurrence is equivalent to the condition $t < \infty$ a.s., i.e. to (i). The equivalence of (i) and (ii) follows from (4.5). □

A recurrent renewal process $(T(i); i \ge 0)$ is called *positive recurrent* if $M_b \overset{\text{def}}{=} \mathbb{E}t = \sum_1^\infty n b_n$ is finite, otherwise *null recurrent.*

For a probabilistic renewal sequence, let

$$B_n = \mathbb{P}\{t > n\} = 1 - b * 1_n.$$

(We use the symbol 1 also to denote the sequence $1 = (1_n; n \ge 0) = (1, 1, \ldots)$.) It follows from the renewal equation (4.3) that

$$B * u = 1. \tag{4.6}$$

Hence in the positive recurrent case the delay distribution e given by

$$e = M_b^{-1} B$$

is an *equilibrium distribution* in the sense that the corresponding delayed renewal sequence $e * u$ is a constant,

$$e * u \equiv M_b^{-1}.$$

Let b again be an arbitrary sequence satisfying the basic hypotheses (4.1). We define the *generating functions* $\hat{b}(r)$ and $\hat{u}(r)$ by

$$\hat{b}(r) = \sum_1^\infty r^n b_n, \quad \hat{u}(r) = \sum_0^\infty r^n u_n, \quad 0 \le r < \infty.$$

Note that $\hat{b}(0) = 0$ and $\hat{u}(0) = 1$. It follows from the renewal equation that we have

$$\hat{u}(r) = 1 + \hat{b}(r)\hat{u}(r) \quad \text{for all } 0 \leq r < \infty,$$

whence

$$\hat{u}(r) = (1 - \hat{b}(r))^{-1} \quad \text{whenever } \hat{b}(r) \leq 1, \tag{4.7a}$$

and

$$\hat{u}(r) = \infty \quad \text{whenever } \hat{b}(r) \geq 1. \tag{4.7b}$$

The real number

$$\begin{aligned} R &= \sup\{r \geq 0 : \hat{u}(r) < \infty\} \\ &= \sup\{r \geq 0 : \hat{b}(r) < 1\} \end{aligned}$$

is called the *convergence parameter* of the renewal sequence u. u is called R-*recurrent* if $\hat{u}(R) = \infty$, otherwise R-*transient*.

Note that, for a probabilistic renewal sequence, $R \geq 1$ and recurrence means the same as 1-recurrence.

From (4.7) and from the monotone convergence theorem we obtain the following:

Proposition 4.3. $\hat{b}(R) \leq 1$ always, and $\hat{b}(R) = 1$ if and only if u is R-recurrent. □

Define a transformed sequence $\tilde{b} = (\tilde{b}_n ; n \geq 0)$ by

$$\tilde{b}_n = R^n b_n.$$

It is easy to see that the renewal sequence $\tilde{u}$ corresponding to $\tilde{b}$ is given by

$$\tilde{u}_n = R^n u_n, \quad n \geq 0. \tag{4.8}$$

From Propositions 4.2 and 4.3 it follows that $\tilde{u}$ is probabilistic, and it is recurrent if and only if u is R-recurrent.

4.2 Kernels and Markov chains having a proper atom

Now we resume our study of non-negative kernels. Before introducing the general regeneration scheme we shall look at the simple special case where the kernel K has a proper atom.

Definition 4.3. A non-empty set $\alpha \in \mathscr{E}$ is called a *proper atom* (for the kernel K), if

(i) $K(x, \cdot) = K(y, \cdot)$ for all $x, y \in \alpha$, and
(ii) $x \to \alpha$ for some $x \in \alpha$.

Let $\alpha \in \mathscr{E}$ be a proper atom. For any function f on E, if f is constant on α, then, by convention, we write $f(\alpha)$ for the common value of $f(x)$, $x \in \alpha$. So, for example, the notation $K(\alpha,\cdot)$ makes sense. Note that, in fact, $K^n(x,\cdot)=K^n(y,\cdot)$ for all $x, y\in\alpha, n\geq 1$, and hence condition (ii) can be written in the form

(ii′) $K^{n_0}(\alpha,\alpha)>0$ for some $n_0\geq 1$ (or just briefly, $\alpha\to\alpha$).

In other words, (ii′) states that the proper atom α is $K^{n_0}(\alpha,\cdot)$-communicating. Consequently, by Proposition 2.3 (ii) the kernel K restricted to the set

$$\alpha^+ = \{x\in E: x\to\alpha\} = \{G1_\alpha > 0\}$$

is irreducible. Since we will be interested only in the restriction of K to α^+, there is usually no loss of generality in assuming that $\alpha^+=E$, i.e. K is irreducible.

From the inequalities $K^{n+1}(\alpha,\cdot)\geq K^n(\alpha,\alpha)K(\alpha,\cdot)$, $n\geq 1$, and since $K^{n+1}(\alpha,\cdot)$ is σ-finite by our Basic assumption, it follows that $K^n(\alpha,\alpha)$ is finite for all $n\geq 1$.

Note that a proper atom is a small set: we have $M(1,1,1_\alpha,\nu)$ with $\nu=K(\alpha,\cdot)$.

Also note that, if $x_0\in E$ is a communicating state, i.e. a state satisfying

$$\{x_0\}\in\mathscr{E} \quad \text{and } x_0\to x_0,$$

then the singleton $\{x_0\}$ is a proper atom.

Suppose now for a while that the kernel K has a proper atom $\alpha\in\mathscr{E}$. It can be shown that with α there is always associated a renewal sequence:

Proposition 4.4. Let $\nu=K(\alpha,\cdot)$.

(i) The sequence u defined by

$$u_n = K^n(\alpha,\alpha)\begin{cases} =1 & \text{for } n=0,\\ =\nu K^{n-1}(\alpha) & \text{for } n\geq 1,\end{cases}$$

is a renewal sequence. Its increment sequence b is given by

$$b_n = K(I_{\alpha^c}K)^{n-1}(\alpha,\alpha) = \nu(I_{\alpha^c}K)^{n-1}(\alpha), \quad n\geq 1.$$

(ii) More generally, for any $x\in E$, the sequence $v(x)$ defined by

$$v_n(x) = K^n(x,\alpha), \quad n\geq 0,$$

is a delayed renewal sequence:

$$v(x) = a(x) * u,$$

where the delay sequence $a(x)$ is given by

$$a_n(x) = (I_{\alpha^c}K)^n(x,\alpha), \quad n\geq 0.$$

(iii) For any $A\in\mathscr{E}$, the sequence $w(A)$ defined by

$$w_n(A) = K^{n+1}(\alpha,A) = \nu K^n(A), \quad n\geq 0,$$

is a delayed renewal sequence:

$$w(A) = u * \sigma(A),$$

where the delay sequence σ(A) is given by

$$\sigma_n(A) = K(I_{\alpha^c}K)^n(\alpha, A) = \nu(I_{\alpha^c}K)^n(A), \quad n \geq 0.$$

In the proof we need the following simple algebraic lemma:

Lemma 4.1. Let γ and δ be two elements of a ring. Then for all $n \geq 1$:

$$(\gamma+\delta)^n = \gamma^n + \sum_{m=1}^{n} \gamma^{m-1}\delta(\gamma+\delta)^{n-m}$$

$$= \gamma^n + \sum_{m=1}^{n} (\gamma+\delta)^{m-1}\delta\gamma^{n-m}.$$

Proof of Lemma 4.1. Write $(\gamma+\delta)^n$ as the sum of γ^n and $2^n - 1$ terms each of which is a product of γs and δs containing at least one δ. In order to obtain the first (resp. the second) of the identities of the lemma, sum together those terms where δ occurs for the first (resp. last) time at the mth position. □

Proof of Proposition 4.4. Set $\gamma = I_{\alpha^c}K$ and $\delta = I_\alpha K$. By the lemma we have for the sequence $v_n(x) = K^n(x, \alpha)$, $n \geq 0$:

$$v_n(x) = (\gamma+\delta)^n(x, \alpha)$$

$$= (I_{\alpha^c}K)^n(x, \alpha) + \sum_{m=1}^{n} (I_{\alpha^c}K)^{m-1} I_\alpha K^{n-m+1}(x, \alpha)$$

$$= a(x) * u_n.$$

Since $u_n = K^n(\alpha, \alpha) = Kv_{n-1}(\alpha)$ and $b_n = K(I_{\alpha^c}K)^{n-1}(\alpha, \alpha) = Ka_{n-1}(\alpha)$, $n \geq 1$, we see that u satisfies the renewal equation $u_n = \delta + b * u_n$, whence by Proposition 4.1 (iii) it is a renewal sequence. Thus the proofs of (i) and (ii) are completed.

The proof of (iii) is analogous to that of (ii). □

We call the sequence $u = (K^n(\alpha, \alpha); n \geq 0)$ the *embedded renewal sequence associated with the proper atom* α.

As a direct consequence of the definitions and of Theorem 3.2 we obtain the following:

Proposition 4.5. The kernel K and the embedded renewal sequence u have a common convergence parameter R. Moreover, K is R-recurrent if and only if u is. □

Next we shall look at the harmonic functions and invariant measures for the kernels rK, $0 < r \leq R$. We retain our basic hypothesis of the existence of a proper atom α.

Let $G_\alpha^{(r)}$ be the potential kernel of $rI_{\alpha^c}K$, i.e.,

$$G_\alpha^{(r)} = \sum_0^\infty r^n (I_{\alpha^c}K)^n, \quad 0 \leq r < \infty,$$

and let $h_\alpha^{(r)} \in \mathscr{E}_+$ and $\pi_\alpha^{(r)} \in \mathscr{M}_+$ be defined by

$$h_\alpha^{(r)}(x) = rG_\alpha^{(r)}(x, \alpha) = \sum_0^\infty r^{n+1}(I_{\alpha^c}K)^n(x, \alpha),$$

$$\pi_\alpha^{(r)}(A) = rKG_\alpha^{(r)}(\alpha, A) = r\nu G_\alpha^{(r)}(A) = \sum_0^\infty r^{n+1}\nu(I_{\alpha^c}K)^n(A).$$

The generating function $\hat{b}(r) = \sum_1^\infty r^n b_n$ of the increment sequence $b = (K(I_{\alpha^c}K)^{n-1}(\alpha, \alpha);\ n \geq 1)$ can be also expressed in the forms

$$\hat{b}(r) = \sum_0^\infty r^{n+1}K(I_{\alpha^c}K)^n(\alpha, \alpha) = r\nu G_\alpha^{(r)}(\alpha)$$
$$= \nu(h_\alpha^{(r)}) = \pi_\alpha^{(r)}(\alpha).$$

It follows from Propositions 4.3 and 4.5 that the convergence parameter R of K equals

$$R = \sup\{r \geq 0 : \nu(h_\alpha^{(r)}) < 1\}.$$

Moreover, $\nu(h_\alpha^{(r)}) \leq 1$, and $\nu(h_\alpha^{(r)}) = 1$ if and only if K is R-recurrent.

From the identity

$$rKh_\alpha^{(r)}(x) = rI_{\alpha^c}KG_\alpha^{(r)}(x, \alpha) + r\nu G_\alpha^{(r)}(x, \alpha)$$
$$= h_\alpha^{(r)}(x) - 1_\alpha(x)(1 - \nu(h_\alpha^{(r)})),$$

we obtain part (i) of the following:

Proposition 4.6. (i) For any $0 < r \leq R$, the function $h_\alpha^{(r)}$ is superharmonic for the kernel rK,

$$h_\alpha^{(r)} \geq rKh_\alpha^{(r)}.$$

If K is R-recurrent, then $h_\alpha^{(R)}$ is harmonic for RK,

$$h_\alpha^{(R)} = RKh_\alpha^{(R)}.$$

(ii) Similarly, the measure $\pi_\alpha^{(r)}$ is subinvariant for rK, that is

$$\pi_\alpha^{(r)} \geq r\pi_\alpha^{(r)}K.$$

If K is R-recurrent then $\pi_\alpha^{(R)}$ is invariant for RK, i.e.,

$$\pi_\alpha^{(R)} = R\pi_\alpha^{(R)}K. \qquad \square$$

In Sections 5.1 and 5.2 we shall make a thorough study of the subinvariant and invariant functions and measures for a general R-recurrent kernel K.

When $K = P$ is the transition probability of a Markov chain (X_n) and P has a proper atom $\alpha \in \mathscr{E}$, we can interpret the situation in probabilistic terms:

Suppose that (X_n) has initial state $X_0 = x \in E$. The epochs $(T_\alpha(i); i \geq 0)$ of the successive visits to the proper atom α form a renewal process corresponding to the incidence process $Y_n = 1_\alpha(X_n), n \geq 0$. Its delay is $T_\alpha(0) = T_\alpha$. The delay distribution $a(x)$ is given by

$$a_n(x) = \mathbb{P}_x\{T_\alpha = n\} = (I_{\alpha^c}P)^n(x, \alpha), \quad n \geq 0.$$

The increments $t_\alpha(i) = T_\alpha(i) - T_\alpha(i-1), i \geq 1$, have the common distribution

$$b_n = \mathbb{P}_\alpha\{S_\alpha = n\} = P(I_{\alpha^c}P)^{n-1}(\alpha, \alpha), \quad n \geq 1.$$

The corresponding undelayed renewal sequence u is

$$u_n = \mathbb{P}_\alpha\{X_n \in \alpha\} = P^n(\alpha, \alpha), \quad n \geq 0,$$

while the delayed renewal sequence $v(x)$ (with start at $X_0 = x$) is given by

$$v_n(x) = \mathbb{P}_x\{X_n \in \alpha\} = P^n(x, \alpha\} = a(x) * u_n, \quad n \geq 0 \tag{4.9}$$

(cf. Proposition 4.4). The identity (4.9) is called the *first entrance* (to the proper atom α) *decomposition* of the transition probability $P^n(x, \alpha)$.

Since $\hat{b}(r) = \sum_0^\infty r^{n+1} P(I_{\alpha^c}P)^n(\alpha, \alpha) = \mathbb{E}_\alpha[r^{S_\alpha}; S_\alpha < \infty]$, we see that the convergence parameter R of P equals

$$R = \sup\{r \geq 0 : \mathbb{E}_\alpha[r^{S_\alpha}; S_\alpha < \infty] < 1\}$$

and P is R-recurrent if and only if

$$\mathbb{E}_\alpha[R^{S_\alpha}; S_\alpha < \infty] = 1.$$

As a special case of Proposition 4.6 we have:

Corollary 4.1. (i) The function h_α defined by

$$h_\alpha(x) = h_\alpha^{(1)}(x) = G_\alpha(x, \alpha) = \mathbb{P}_x\{T_\alpha < \infty\}$$

is superharmonic, i.e., $h_\alpha \geq Ph_\alpha$. h_α is harmonic, $h_\alpha = Ph_\alpha$, if (X_n) is recurrent.

(ii) The measure π_α defined by

$$\pi_\alpha(A) = \pi_\alpha^{(1)}(A) = \nu G_\alpha(A) = \mathbb{E}_\alpha \sum_1^{S_\alpha} 1_A(X_n)$$

is subinvariant, i.e., $\pi_\alpha \geq \pi_\alpha P$. π_α is invariant, $\pi_\alpha = \pi_\alpha P$, if (X_n) is recurrent. □

As a converse result to Proposition 4.4 (i), we can easily show that an arbitrary undelayed renewal process $(T(i); i \geq 0)$ can be obtained as the embedded renewal process of a Markov chain:

Let $\bar{M} = \operatorname{ess\,sup} t = \sup\{m \geq 1 : b_m > 0\}$. If $\bar{M}$ is finite, set $E =$

$\{0,1,\ldots,\bar{M}-1\}$; otherwise set $E=\mathbb{N}$. Define the transition matrix $P=(p(x,y);\ x,y\in E)$ by

$$p(x,x+1)=\mathbb{P}\{t>x+1\,|\,t>x\}=B_x^{-1}B_{x+1},$$
$$p(x,0)=\mathbb{P}\{t=x+1\,|\,t>x\}=B_x^{-1}b_{x+1},$$
$$p(x,y)=0\quad\text{otherwise.}$$

Now one observes easily that the above transition probabilities are the transition probabilities of the *backward chain* $(V_n;n\geq 0)$ of the undelayed renewal process $(T(i);i\geq 0)$,

$$V_n\stackrel{\text{def}}{=}n-\max\{m:0\leq m\leq n,\ Y_m=1\},\quad n\geq 0,\tag{4.10}$$

i.e., V_n counts the time elapsed since the last renewal epoch before (or at) n. The renewal process $(T(i);i\geq 0)$ is reobtained as the embedded renewal process associated with the proper atom $\alpha=\{0\}$.

Examples 4.1. (*a*) For a Card_F-irreducible matrix $K=(k(x,y);\ x,y\in E)$ every state $z\in F$ is communicating (whence $\{z\}$ is a proper atom).

For any $z\in F$, let ${}_zK=({}_zk(x,y);x,y\in E)$ denote the matrix obtained by removing the zth row:

$$\begin{aligned}{}_zk(x,y)&=k(x,y)-1_{\{z\}}(x)k(x,y)\\&=\begin{cases}k(x,y) & \text{for}\quad x\neq z,\\ 0 & \text{for}\quad x=z.\end{cases}\end{aligned}$$

Let $({}_zK)^n=({}_zk^{(n)}(x,y);x,y\in E)$.

If K is Card_F-irreducible and R-recurrent then the column vector h_z defined by

$$h_z(x)=\sum_{n=0}^{\infty}R^n{}_zk^{(n)}(x,z),\quad x\in E,$$

is the unique non-negative column vector satisfying

$$RKh_z=h_z\quad\text{and}\quad h_z(z)=1.$$

If $K=P$ is the transition matrix of a Card_F-irreducible, discrete Markov chain then for all $z\in F$ (writing $S_z=S_{\{z\}}$):

$$\mathbb{E}_z[r^{S_z};S_z<\infty]\begin{cases}\leq 1 & \text{for all}\quad r\leq R,\\ >1 & \text{for all}\quad r>R,\end{cases}$$

and P is R-recurrent if and only if equality holds for $r=R$. In the R-recurrent case the R-invariant column vector h_z is given by

$$h_z(x)=\mathbb{E}_x[R^{T_z};T_z<\infty]=\mathbb{E}_x[R^{S_z};S_z<\infty].$$

(*d*) (The reflected random walk). If $F(0)=\mathbb{P}\{z_1\leq 0\}>0$, then the state 0 is communicating (whence $\{0\}$ is a proper atom).

The convergence parameter R is given by

$$R = \sup\{r \geq 0 : \mathbb{E}_0[r^{S_0}; S_0 < \infty] < 1\}.$$

The reflected random walk (W_n) is R-recurrent if and only if $\mathbb{E}_0[R^{S_0}; S_0 < \infty] = 1$; the unique R-invariant function h_0 satisfying $h_0(0) = 1$ is given by

$$h_0(x) = \mathbb{E}_x[R^{T_0}; T_0 < \infty], \quad x \in \mathbb{R}_+.$$

Suppose that (W_n) is recurrent. Then the measure

$$\pi_0(A) = \mathbb{E}_0\left[\sum_1^{S_0} 1_A(X_n)\right], \quad A \in \mathscr{R}_+,$$

is the unique invariant measure satisfying $\pi_0(\{0\}) = 1$.

In addition to our previous examples (*a*)–(*h*) we shall introduce two new examples:

(*i*) Consider the following model for a *storage*:

Let $0 \leq \underline{M} < \bar{M} < \infty$ be two constants (the lower and upper level, respectively, of the storage). Let z_n, $n = 1, 2, \ldots$, be i.i.d., non-negative random variables (the daily demands). Let S_0 be a random variable, independent of $\{z_n ; n \geq 1\}$ (the initial storage). We suppose that the daily storages $S_n, n \geq 0$, are obtained as follows: Whenever S_{n-1}, the storage at day $n-1$, is above the lower level $\underline{M}$ the storage decreases by the daily demand z_n, whereas if the storage S_{n-1} is below $\underline{M}$ then the storage will be (immediately) supplied to the upper level $\bar{M}$ (from which it again decreases by the amount z_n); i.e.,

$$\begin{aligned} S_n &= S_{n-1} - z_n \quad \text{if} \quad S_{n-1} > \underline{M}, \\ &= \bar{M} - z_n \qquad \text{if} \quad S_{n-1} \leq \underline{M}. \end{aligned}$$

Clearly $(S_n ; n \geq 0)$ is a Markov chain on $(\mathbb{R}, \mathscr{R})$. The half-interval $(-\infty, \bar{M}]$ is an absorbing set. The half-interval $\alpha = (-\infty, \underline{M}]$ is a proper atom. The probability measure $\nu = P(\alpha, \cdot)$ is given by

$$\nu = \mathscr{L}(\bar{M} - z_1).$$

(*j*) The waiting times of successive customers in a 2-*server queue* can be modelled as follows:

Let $t_1, t_2, \ldots$ be i.i.d., non-negative random variables (the interarrival times), and let $s_0, s_1, \ldots$ be i.i.d., non-negative random variables, independent of $\{t_n ; n \geq 1\}$ (the service times). Let $E = \{x = (x^{(1)}, x^{(2)}) \in \mathbb{R}^2 : 0 \leq x^{(1)} \leq x^{(2)}\}$, and let $W_0 = (W_0^{(1)}, W_0^{(2)})$ be an E-valued random element, i.e.,

$$0 \leq W_0^{(1)} \leq W_0^{(2)} \quad \text{(a.s.)}.$$

We define iteratively a Markov chain $(W_n ; n \geq 0) = (W_n^{(1)}, W_n^{(2)} ; n \geq 0)$ on E:

$$W_n^{(1)} = \min\{(W_{n-1}^{(1)} + s_{n-1} - t_n)_+, (W_{n-1}^{(2)} - t_n)_+\},$$
$$W_n^{(2)} = \max\{(W_{n-1}^{(1)} + s_{n-1} - t_n)_+, (W_{n-1}^{(2)} - t_n)_+\}.$$

The component $W_n^{(1)}$ (which, in general, is not a Markov chain) represents the waiting time of the nth customer (before service).

Let $\bar{t} = \operatorname{ess\,sup} t_1, \underline{s} = \operatorname{ess\,inf} s_0$. If

$$\underline{s} < \bar{t}$$

then $(0, 0) \in E$ is a communicating state. (Note that, if

$$\underline{s} > 2\bar{t}$$

then (W_n) is not even irreducible. We will look at the case $\bar{t} < \underline{s} < 2\bar{t}$ later in Example 4.2(j).)

4.3 The general regeneration scheme

Our next object is to show that a regeneration scheme similar to that described above exists for an arbitrary irreducible kernel satisfying the minorization condition $M(m_0, 1, s, \nu)$ with $m_0 = 1$.

We assume that K is irreducible.

Definition 4.4. A pair (s, ν), $s \in \mathscr{E}^+$, $\nu \in \mathscr{M}^+$, is called an *atom* (for K) if the minorization condition $M(1, 1, s, \nu)$ holds, i.e.,

$$K \geq s \otimes \nu.$$

Concerning the irreducibility assumption similar remarks could be made here as were made after Definition 4.3. So, instead of assuming that K is irreducible and $s \in \mathscr{E}^+$, it would be sufficient to assume only that s is attainable from ν, i.e.,

$$\nu K^n s > 0 \quad \text{for some } n \geq 0.$$

But this would imply irreducibility: the restriction of K to the set $\{s > 0\}^+ = \{Gs > 0\}$ would be irreducible.

Note that, if we have $M(m_0, \beta, s, \nu)$ with $m_0 \geq 1, \beta > 0$ arbitrary, then the pair $(\beta s, \nu)$ is an atom for the iterated kernel K^{m_0}.

Note also that, if $\alpha \in \mathscr{E}^+$ is a proper atom then the pair $(1_\alpha, K(\alpha, \cdot))$ is an atom.

In Proposition 4.4 we proved that with a proper atom there is associated a renewal sequence. It turns out that this holds true also in the case where K has only an atom (s, ν):

Theorem 4.1. Let (s, ν) be an atom. Then:

(i) The sequence $u = (u_n; n \geq 0)$ defined by

$$u_0 = 1, \quad u_n = \nu K^{n-1} s, \quad n \geq 1,$$

is a renewal sequence. Its increment sequence b is given by

$$b_0 = 0, \quad b_n = v(K - s \otimes v)^{n-1} s, \quad n \geq 1.$$

(ii) For any $x \in E$, the sequence $(K^n s(x); n \geq 0)$ is a delayed renewal sequence. Its delay sequence $a(x) = (a_n(x); n \geq 0)$ is given by

$$a_n(x) = (K - s \otimes v)^n s(x), \quad n \geq 0; \tag{4.11}$$

i.e., we have

$$K^n s(x) = a(x) * u_n, \quad n \geq 0.$$

(iii) For any $A \in \mathscr{E}$, the sequence $(vK^n(A); n \geq 0)$ is a delayed renewal sequence. Its delay sequence $\sigma(A) = (\sigma_n(A); n \geq 0)$ is given by

$$\sigma_n(A) = v(K - s \otimes v)^n(A), \quad n \geq 0;$$

i.e., we have

$$vK^n(A) = u * \sigma(A)_n, \quad n \geq 0.$$

(iv) For all $n \geq 1, x \in E, A \in \mathscr{E}$:

$$K^n(x, A) = (K - s \otimes v)^n(x, A) + a(x) * u * \sigma(A)_{n-1}.$$

Proof. (i) and (ii): Set $\gamma = K - s \otimes v$, $\delta = s \otimes v$ in Lemma 4.1. We get for all $n \geq 1$:

$$K^n = (K - s \otimes v)^n + \sum_{m=1}^{n} (K - s \otimes v)^{m-1} s \otimes vK^{n-m}, \tag{4.12}$$

Therefore

$$K^n s(x) = \sum_{m=0}^{n} (K - s \otimes v)^m s(x) u_{n-m} = a(x) * u_n.$$

Since $b_n = v(a_{n-1})$ for all $n \geq 1$, it follows that u satisfies the renewal equation $u = \delta + b * u$. Hence it is a renewal sequence.

(iii) The proof is analogous to that of (ii).

(iv) By (4.12) and (iii),

$$\begin{aligned} K^n(x, A) &= (K - s \otimes v)^n(x, A) + \sum_{m=1}^{n} a_{m-1}(x) u * \sigma(A)_{n-m} \\ &= (K - s \otimes v)^n(x, A) + a(x) * u * \sigma(A)_{n-1}. \qquad \square \end{aligned}$$

We write $G^{(r)}_{s,v}$ for the potential kernel of the kernel $r(K - s \otimes v)$, i.e.,

$$G^{(r)}_{s,v} = \sum_{0}^{\infty} r^n (K - s \otimes v)^n, \quad r \geq 0. \tag{4.13}$$

Recall that $G^{(r)}$ denotes the potential kernel of rK, $G^{(r)} = \sum_0^\infty r^n K^n$.

For the generating functions $\hat{u}(r)=\sum_0^\infty r^n u_n$ and $\hat{b}(r)=\sum_1^\infty r^n b_n$ we obtain

$$\hat{u}(r)=1+\sum_1^\infty r^n vK^{n-1}s=1+rvG^{(r)}s,$$

$$\hat{b}(r)=\sum_1^\infty r^n v(K-s\otimes v)^{n-1}s=rvG^{(r)}_{s,v}s.$$

As a direct consequence of Theorem 3.2 and Proposition 4.3 we have:

Proposition 4.7. Suppose that K is an irreducible kernel and that it has an atom (s,v). Then:

(i) The convergence parameter R of the kernel K and of the embedded renewal sequence u coincide, and

$$R=\sup\{r\geq 0:\hat{u}(r)<\infty\}=\sup\{r\geq 0:\hat{b}(r)<1\};$$

(ii) $\hat{b}(R)\leq 1$ always;

(iii) K R-recurrent $\Leftrightarrow \hat{u}(R)=\infty \Leftrightarrow \hat{b}(R)=1$. □

For later purposes note the following formula: Suppose that K is 1-recurrent. Then the embedded renewal sequence u is recurrent, i.e., $M_b^{(0)}=\hat{b}(1)=1$, and, writing $h=G^{(1)}_{s,v}s$, we have

$$B_n \stackrel{\text{def}}{=} 1-b*1_n=\sum_{n+1}^\infty b_m=\sum_n^\infty v(K-s\otimes v)^m s=\sigma_n(h). \tag{4.14}$$

4.4 The split chain

When $K=P$ is a transition probability the results of Section 4.3 have an interesting probabilistic interpretation. We will show in Theorem 4.2 that an atom (s,v) for P in fact represents a proper atom in a suitably enlarged state space.

Suppose now for a while that $K=P$ is the transition probability of an irreducible Markov chain $(X_n;n\geq 0)$. We suppose that (s,v) is an atom, i.e., P satisfies the minorization condition $M(m_0,1,s,v)$ with $m_0=1$,

$$P\geq s\otimes v.$$

We can assume without loss of generality that v is a probability measure (cf. Remark 2.1(v)). Then necessarily $0\leq s\leq 1$.

Define a kernel Q on $(E,\mathscr{E})$ by setting

$$\begin{aligned} Q(x,A)&=(1-s(x))^{-1}(P(x,A)-s(x)v(A)) \quad \text{if } s(x)<1,\\ &=1_A(x) \quad \text{if } s(x)=1. \end{aligned}$$

Then Q is clearly substochastic. The transition probability P splits into two parts:

$$P(x,A)=s(x)v(A)+(1-s(x))Q(x,A).$$

Thus a transition starting from an arbitrary state $x \in E$ can be interpreted as happening in two stages. First, an $s(x)$-coin (that is a coin with $\mathbb{P}\{\text{'head'}\} = s(x)$) is tossed. If the result is 'head' the Markov chain moves according to the probability law $\nu(\cdot)$, otherwise according to $Q(x, \cdot)$. The crucial point here is the fact that the occurrence of 'head' leads to a transition law which is independent of the state x. In what follows we shall formulate this heuristic approach in precise terms by adjoining to the Markov chain (X_n) a $\{0,1\}$-valued stochastic process (Y_n) representing precisely the successive results of the tossing of an $s(X_n)$-coin at $n = 0, 1, \ldots$. It is obvious that we thus obtain a Markov chain $(X_n, Y_n;\ n \geq 0)$ on the state space $E \times \{0,1\} \cong E \times \{\text{'tail', 'head'}\}$ such that the subset $E \times \{1\} \cong E \times \{\text{'head'}\}$ is a proper atom for (X_n, Y_n). Hence there exists an embedded renewal process, the incidence process of which is the sequence (Y_n).

To formalize all this, suppose that the Markov chain $(X_n; n \geq 0)$ is Markov w.r.t. an arbitrary, fixed history $(\mathscr{F}_n)$. If the transition probability P is not stochastic we complete it to a stochastic kernel by extending it to $(E_\Delta, \mathscr{E}_\Delta)$ as described in Section 1.2. If necessary, we enlarge the underlying probability space $(\Omega, \mathscr{F})$ into the product space $(\Omega \times \{0,1\}^{\times\infty}, \mathscr{F} \otimes \{\varnothing, \{0\}, \{1\}, \{0,1\}\}^{\otimes\infty})$, to include the results of a coin tossing experiment.

Let $(Y_n; n \geq 0)$ be a $\{0,1\}$-valued stochastic process depending on the Markov chain $(X_n, \mathscr{F}_n)$ through the formulas

$$\begin{aligned}\mathbb{P}\{X_{n+1} \in A,\ Y_n = 1 \mid \mathscr{F}_n \vee \mathscr{F}^Y_{n-1}\} &= \mathbb{P}\{X_{n+1} \in A,\ Y_n = 1 \mid X_n\}\\ &= s(X_n)\nu(A),\end{aligned} \tag{4.15a}$$

$$\begin{aligned}\mathbb{P}\{X_{n+1} \in A,\ Y_n = 0 \mid \mathscr{F}_n \vee \mathscr{F}^Y_{n-1}\} &= \mathbb{P}\{X_{n+1} \in A,\ Y_n = 0 \mid X_n\}\\ &= P(X_n, A) - s(X_n)\nu(A),\end{aligned} \tag{4.15b}$$

$A \in \mathscr{E}_\Delta, n \geq 0$. (By convention $\mathscr{F}^Y_{-1} = \{\varnothing, \Omega\}$ and $\nu(\{\Delta\}) = 0$.) Note that the conditions (4.15*a*) and (4.15*b*) are together equivalent to the following set of conditions:

$$\mathbb{P}\{X_{n+1} \in A \mid \mathscr{F}_n \vee \mathscr{F}^Y_{n-1}\} = \mathbb{P}\{X_{n+1} \in A \mid X_n\} = P(X_n, A), \tag{4.16a}$$

$$\mathbb{P}\{Y_n = 1 \mid \mathscr{F}_n \vee \mathscr{F}^Y_{n-1}\} = \mathbb{P}\{Y_n = 1 \mid X_n\} = s(X_n), \tag{4.16b}$$

$$\mathbb{P}\{X_{n+1} \in A \mid \mathscr{F}_n \vee \mathscr{F}^Y_{n-1}; Y_n = 1\} = \mathbb{P}\{X_{n+1} \in A \mid Y_n = 1\} = \nu(A), \tag{4.16c}$$

$A \in \mathscr{E}_\Delta, n \geq 0$. Condition (4.16*a*) simply states that (X_n) is Markov w.r.t. the history $(\mathscr{F}_n \vee \mathscr{F}^Y_{n-1})$. Condition (4.16*b*) means that the probability of getting a 'head' at the nth toss is equal to $s(X_n)$ independently of the previous history $\mathscr{F}_n$ of the chain (X_n) and of the tosses up to $n-1$. Condition (4.16c) says that, if 'head' is obtained at the nth toss then the next transition obeys the probability law $v(\cdot)$ independently of the past history $\mathscr{F}_n$ of the chain and of the tosses up to $n-1$.

From conditions (4.16a, b, c) it is also easy to derive the formula

$$\mathbb{P}\{X_{n+1}\in A \mid \mathscr{F}_n \vee \mathscr{F}^Y_{n-1}; Y_n = 0\} = Q(X_n, A), \tag{4.16d}$$

which has a similar interpretation to the formulas above.

Note also that

$$\begin{aligned}\mathbb{P}\{Y_n = 1 \mid \mathscr{F}^X \vee \mathscr{F}^Y_{n-1}\} &= \mathbb{P}\{Y_n = 1 \mid X_n, X_{n+1}\}\\ &= r(X_n, X_{n+1}),\end{aligned} \tag{4.17}$$

where r is defined as the Radon–Nikodym derivative

$$r(x, y) = \frac{s(x)v(\mathrm{d}y)}{P(x, \mathrm{d}y)}.$$

In view of the discussion above, the following result is not surprising:

Theorem 4.2. (i) The stochastic process $(X_n, Y_n;\ n \geq 0)$ is a Markov chain w.r.t. the history $(\mathscr{F}_n \vee \mathscr{F}^Y_n;\ n \geq 0)$.

(ii) The Markov chain $(X_n, Y_n;\ n \geq 0)$ is irreducible.

(iii) The set $\alpha \stackrel{\text{def}}{=} E \times \{1\}$ is a proper atom for (X_n, Y_n).

(iv) The renewal sequence

$$u_0 = 1, \quad u_n = vP^{n-1}s, \quad n \geq 1,$$

associated with the atom (s, v) of the kernel P is also the renewal sequence associated with the proper atom α of the Markov chain (X_n, Y_n).

Proof. It follows from (4.15) and (4.16) that

$$\begin{aligned}\mathbb{P}\{X_{n+1}\in \mathrm{d}y,\ Y_{n+1} = 1 \mid \mathscr{F}_n \vee \mathscr{F}^Y_n\}\\ &= \mathbb{P}\{X_{n+1}\in \mathrm{d}y,\ Y_{n+1} = 1 \mid X_n, Y_n\}\\ &= \begin{cases} v(\mathrm{d}y)s(y) & \text{if } Y_n = 1,\\ Q(X_n, \mathrm{d}y)s(y) & \text{if } Y_n = 0,\end{cases}\end{aligned}$$

and

$$\begin{aligned}\mathbb{P}\{X_{n+1}\in \mathrm{d}y,\ Y_{n+1} = 0 \mid \mathscr{F}_n \vee \mathscr{F}^Y_n\}\\ &= \mathbb{P}\{X_{n+1}\in \mathrm{d}y,\ Y_{n+1} = 0 \mid X_n, Y_n\}\\ &= \begin{cases} v(\mathrm{d}y)(1 - s(y)) & \text{if } Y_n = 1,\\ Q(X_n, \mathrm{d}y)(1 - s(y)) & \text{if } Y_n = 0.\end{cases}\end{aligned}$$

Consequently, (X_n, Y_n) is a Markov chain w.r.t. the history $(\mathscr{F}_n \vee \mathscr{F}^Y_n)$. We also see that the transitions starting from $\alpha = E \times \{1\}$ are identical in distribution.

In order to show that the chain (X_n, Y_n) is irreducible and that α is a proper atom for it, it suffices to show that

$$\mathbb{P}\{Y_n = 1 \quad \text{for some } n \geq 1 \mid X_0 = x, Y_0 = i\} > 0$$

for any initial state $(x, i) \in E \times \{0, 1\}$ of the bivariate chain (X_n, Y_n). By (4.16)

$$\mathbb{P}\{Y_n = 1 \mid X_0 = x, Y_0 = 1\} = \nu P^{n-1} s \quad \text{for all } n \geq 1. \tag{4.18}$$

By irreducibility $\nu P^{n-1} s > 0$ for some $n > 1$. Similarly,

$$\mathbb{P}\{Y_n = 1 \mid X_0 = x, Y_0 = 0\} = \int Q(x, \mathrm{d}y) P^{n-1} s(y)$$
$$> 0 \quad \text{for some } n \geq 1.$$

The result (4.18) also proves (iv). □

We call the bivariate Markov chain $(X_n, Y_n;\ n \geq 0)$ the *split chain* of (X_n). We write $\mathbb{P}_\alpha$ for the probability measure defined on the σ-algebra $\sigma(X_n, n \geq 1;\ Y_n, n \geq 0)$ and corresponding to the initial state $Y_0 = 1$ ($X_0 = x$ arbitrary), i.e. $\mathbb{P}_\alpha = \mathscr{L}(X_n, n \geq 1;\ Y_n, n \geq 0 \mid Y_0 = 1)$. By (4.16)

$$\mathbb{E}_\alpha[\zeta \circ \theta] = \mathbb{E}_\nu[\zeta]$$

for any non-negative functional ζ of the split chain (X_n, Y_n). In what follows we shall use the symbol P also to denote the transition probability of the split chain. So, for example, we can write

$$P^n(\alpha, A) = \mathbb{P}_\alpha\{X_n \in A\} = \nu P^{n-1}(A), \quad A \in \mathscr{E}_\Delta, n \geq 1,$$
$$P^n(x, \alpha) = \mathbb{P}_x\{Y_n = 1\} = \mathbb{E}_x s(X_n)$$
$$= P^n s(x), \quad x \in E, n \geq 0,$$
$$P^n(\alpha, \alpha) = \mathbb{P}_\alpha\{Y_n = 1\} = u_n = \begin{cases} 1, & n = 0, \\ \nu P^{n-1} s, & n \geq 1. \end{cases} \tag{4.19}$$

Define the stopping times S_α, T_α and $T_\alpha(i)$, $i \geq 0$, for the split chain (X_n, Y_n) as the hitting times of the proper atom α,

$$S_\alpha = \inf\{n \geq 1 \colon Y_n = 1\},$$
$$T_\alpha = T_\alpha(0) = \inf\{n \geq 0 \colon Y_n = 1\},$$

and iteratively for $i \geq 1$,

$$T_\alpha(i) = \inf\{n > T_\alpha(i-1) \colon Y_n = 1\}.$$

The sequence $(T_\alpha(i); i \geq 0)$ is the embedded renewal process associated with the proper atom α of the split chain. Its increment distribution $b = (b_n;\ n \geq 1)$ is given by

$$b_n = \mathbb{P}_\alpha\{S_\alpha = n\} = \nu(P - s \otimes \nu)^{n-1} s, \quad n \geq 1, \tag{4.20}$$

and the delay distribution $a(x)=(a_n(x);n\geq 0)$ corresponding to the start $X_0=x$ by

$$a_n(x)=\mathbb{P}_x\{T_\alpha=n\}=(P-s\otimes v)^n s(x), \quad n\geq 0. \tag{4.21}$$

Thus the decomposition of Theorem 4.1(ii) for $K=P$ can be interpreted as the *first entrance decomposition*

$$\begin{aligned} P^n s(x) &= \mathbb{P}_x\{Y_n=1\} \\ &= \sum_{m=0}^{n} \mathbb{P}_x\{T_\alpha=m\}\mathbb{P}_\alpha\{Y_{n-m}=1\} \\ &= a(x)*u_n,\ n\geq 0, \end{aligned}$$

(cf. also (4.9)). Similarly, writing

$$\begin{aligned} \sigma_n(A) &= v(P-s\otimes v)^n(A) \\ &= \mathbb{P}_\alpha\{X_{n+1}\in A,\ S_\alpha\geq n+1\}, \quad n\geq 0, \end{aligned} \tag{4.22}$$

$$L_\alpha(n)=\max\{m:\ Y_m=1,\ 0\leq m\leq n\}, \quad n\geq 0,$$

we can interpret part (iii) of Theorem 4.1 as the *last exit decomposition*

$$\begin{aligned} vP^{n-1}(A) &= \mathbb{P}_\alpha\{X_n\in A\} \\ &= \sum_{m=0}^{n-1} \mathbb{P}_\alpha\{L_\alpha(n-1)=m,\ X_n\in A\} \\ &= \sum_{m=0}^{n-1} \mathbb{P}_\alpha\{Y_m=1\}\mathbb{P}_\alpha\{X_{n-m}\in A,\ S_\alpha\geq n-m\} \\ &= u*\sigma(A)_{n-1}, \quad n\geq 1, \end{aligned} \tag{4.23}$$

and part (iv) as the *first-entrance–last-exit decomposition*

$$\begin{aligned} P^n(x,A) &= \mathbb{P}_x\{X_n\in A\} \\ &= \mathbb{P}_x\{X_n\in A,\ T_\alpha\geq n\} \\ &\quad + \sum_{m=0}^{n-1}\sum_{p=m}^{n-1} \mathbb{P}_x\{T_\alpha=m,\ L_\alpha(n-1)=p,\ X_n\in A\} \\ &= (P-s\otimes v)^n(x,A)+a(x)*u*\sigma(A)_{n-1}, \quad n\geq 1. \end{aligned} \tag{4.24}$$

Note that, by (4.17), $T_\alpha+1$ is a randomized stopping time for (X_n); we have

$$\mathbb{P}\{T_\alpha+1=n\,|\,\mathscr{F}^X;\ T_\alpha+1\geq n\}=r(X_{n-1},X_n), \quad n\geq 1. \tag{4.25}$$

Set $G_\alpha=G_{s,v}^{(1)}=\sum_0^\infty (P-s\otimes v)^n$. Note that by (4.19) and (4.20),

$$\sum_0^\infty u_n = 1+vGs\leq\infty,$$

$$\sum_1^\infty b_n=\mathbb{P}_\alpha\{S_\alpha<\infty\}=vG_\alpha s\leq 1.$$

As a direct consequence of Proposition 4.7 we have the following characterization for the recurrence of the Markov chain (X_n):

Corollary 4.2. Suppose that (X_n) is an irreducible Markov chain having an atom (s, ν). Then: (X_n) is recurrent $\Leftrightarrow$ the embedded renewal process $(T_\alpha(i); i \geq 0)$ is recurrent $\Leftrightarrow \nu G s = \infty \Leftrightarrow \nu G_\alpha s = 1$. □

Let us write

$$h_\alpha(x) = \mathbb{P}_x\{T_\alpha < \infty\} = G_\alpha s(x) \quad \text{by (4.21)},$$
$$h_\alpha^\infty(x) = \mathbb{P}_x\{Y_n = 1 \quad \text{i.o.}\}.$$

By the Markov property, and since the (undelayed) renewal process $(T_\alpha(i); i \geq 0)$ is either recurrent or transient, we have

$$h_\alpha^\infty \equiv 0 \text{ in the transient case,}$$
$$h_\alpha^\infty = h_\alpha \text{ in the recurrent case.}$$

In the latter case, let

$$H_\alpha = \{h_\alpha^\infty = 1\} = \{h_\alpha = 1\}.$$

For the following proposition recall from Proposition 3.13 the existence of the harmonic function $\underline{h}$, which is the minimal superharmonic function satisfying $h = 1$ ψ-a.e.

Proposition 4.8. Suppose that (X_n) is a recurrent Markov chain having an atom (s, ν). Then $h_\alpha = \underline{h}$ and $H_\alpha = \bar{H}$.

In particular, (X_n) is Harris recurrent if and only if $h_\alpha \equiv 1$.

Proof. Note that by Corollary 4.2, $\nu(h_\alpha) = 1$. It follows that

$$h_\alpha = G_\alpha s = s + (P - s \otimes \nu) h_\alpha = P h_\alpha,$$

i.e., h_α is harmonic.

In order to prove the minimality of h_α, take any superharmonic function $h \in \mathscr{E}_+$ such that $h = 1$ ψ-a.e. Then

$$h \geq Ph = s + (P - s \otimes \nu) h, \text{ since } \nu \ll \psi.$$

Now apply Theorem 3.1(iii) with $f = s$, $K = P - s \otimes \nu$ to prove that $h \geq h_\alpha$. Hence h_α is minimal. That the set $H_\alpha = \{h_\alpha = 1\}$ is the maximal Harris set follows from Proposition 3.13(iii). □

Apart from the coin tossing experiment, there is an alternative interpretation for our basic construction. For this we need the following:

Definition 4.5. A randomized stopping time T (for (X_n)), satisfying

$$\mathbb{P}_x\{T < \infty\} > 0 \quad \text{for all } x \in E,$$

is called a *regeneration time* (for (X_n)), if there is probability measure φ on $(E, \mathscr{E})$ such that

$$\mathscr{L}(X_n | \mathscr{F}^X_{n-1}; T=n) = \varphi \quad \text{for all } n \geq 0$$

(by convention, $\mathscr{F}^X_{-1} = \{\varnothing, \Omega\}$).

If there exists a regeneration time, we say that the Markov chain (X_n) is *regenerative*. The measure φ is called the regeneration measure.

Note that, if (X_n) is regenerative, then it is φ-irreducible.

According to the following theorem (X_n) is regenerative if and only if the transition probability P has an atom.

If P has an atom (s, ν), we let r denote a version, $r \in \mathscr{E} \otimes \mathscr{E}$, $0 \leq r \leq 1$, of the Radon–Nikodym derivative

$$r(x, y) = \frac{s(x)\nu(\mathrm{d}y)}{P(x, \mathrm{d}y)}.$$

Let $T_{s,\nu}$ denote the randomized stopping time defined by

$$\mathbb{P}\{T_{s,\nu} = n | \mathscr{F}^X; T_{s,\nu} \geq n\} = r(X_{n-1}, X_n), \quad n \geq 1.$$

Note that by (4.25),

$$T_{s,\nu} \stackrel{\mathscr{L}}{=} T_\alpha + 1. \tag{4.26}$$

Theorem 4.3. Suppose that the Markov chain (X_n) is irreducible.

(i) If the transition probability P has an atom (s, ν) then the randomized stopping time $T_{s,\nu}$ is a regeneration time. The corresponding regeneration measure φ is equal to ν.

(ii) Conversely, suppose that (X_n) is regenerative with regeneration measure φ. Then there is a small function $s \in \mathscr{S}^+$ such that the pair (s, φ) is an atom for P.

Proof. (i) The assertion follows directly from (4.16*c*) and (4.26)

(ii) Let T be a regeneration time for (X_n) with corresponding regeneration measure φ. We have for any $n \geq 1$:

$$\begin{aligned} P(x, \mathrm{d}y) &\geq \mathbb{P}\{X_n \in \mathrm{d}y, T = n | \mathscr{F}^X_{n-1}; X_{n-1} = x\} \\ &= \mathbb{P}\{T = n | \mathscr{F}^X_{n-1}; X_{n-1} = x\}\varphi(\mathrm{d}y). \end{aligned}$$

Let

$$s(x) = \sup_{n \geq 1} \{\text{ess sup } \mathbb{P}\{T = n | \mathscr{F}^X_{n-1}; X_{n-1} = x\}\}.$$

Then $P \geq s \otimes \varphi$. If it were true that $s = 0$ ψ-a.e., then this would imply that $\mathbb{P}_\psi\{T < \infty\} = 0$ – a contradiction. Consequently, (s, φ) is an atom for P. □

Example 4.2. (*j*) Consider the 2-server queueing chain (W_n) introduced in Example 4.1(*j*). Suppose that

$$\bar{t} < \underline{s} < 2\bar{t}.$$

Then the subset $F = \{x \in E : x^{(2)} \geq \underline{s} - t\}$ is an absorbing set for (W_n), and (0, 0) is not a communicating state.

However, (W_n) is still regenerative. To see this, let γ be any constant satisfying

$$\underline{s} - \bar{t} < \gamma < \bar{t}.$$

Then the randomized stopping time T_γ defined by

$$T_\gamma = \inf\{n \geq 1 : W_{n-1}^{(1)} = 0,\ W_{n-1}^{(2)} \leq \gamma,\ t_n \geq \gamma\}$$

is a regeneration time for (W_n). The corresponding regeneration measure φ_γ is given by

$$\varphi_\gamma = \mathscr{L}((0, (s_0 - t_1^{(\gamma)})_+)),$$

where $t_1^{(\gamma)}$ is a random variable, independent of s_0, and having the same distribution as t_1 given $t_1 \geq \gamma$; i.e., at $T_\gamma = n$

$$W_n^{(1)} = 0 \quad \text{a.s.},$$

and

$$\begin{aligned} W_n^{(2)} &\overset{\mathscr{L}}{=} (s_{n-1} - t_n)_+ \quad \text{given } t_n \geq \gamma, \\ &\overset{\mathscr{L}}{=} (s_0 - t_1)_+ \quad \text{given } t_1 \geq \gamma, \\ &\overset{\mathscr{L}}{=} (s_0 - t_1^{(\gamma)})_+. \end{aligned}$$

5

Positive and null recurrence

Our interest in Chapter 5 concentrates on the existence and uniqueness of R-invariant functions and measures. (In Proposition 4.6 we have already briefly dealt with this problem under the assumption that K has a proper atom.)

The main result of this chapter states that an irreducible R-recurrent kernel K always has an (essentially) unique R-invariant function h and an (essentially) unique R-invariant measure π. This gives rise to a classification of R-recurrent kernels: If $\pi(h)$ is finite we call the kernel K R-positive recurrent, otherwise R-null recurrent.

In the case where $K = P$ is the transition probability of a recurrent Markov chain, positive recurrence (i.e. $\pi(E) < \infty$) means probabilistically that the chain has a stationary probability distribution. This can be achieved by normalizing π to a probability measure. Then, given that $\mathscr{L}(X_0) = \pi$,

$$\mathscr{L}(X_n) = \pi \quad \text{for all } n \geq 1.$$

It is easy to see that, in fact, given $\mathscr{L}(X_0) = \pi$, the Markov chain $(X_n; n \geq 0)$ is a stationary process.

We shall classify positive recurrent Markov chains further by looking at the 'speeds' with which the chain returns to the π-positive sets in the stationary situation. Later in Chapter 6 we will see that these 'rates of recurrence' correspond to the rates with which the n-step transition probabilities $P^n(x, A)$ tend to their stationary limits $\pi(A)$.

By using a 'similarity transformation' which transforms an R-recurrent kernel into a Harris recurrent transition probability we can extend the rate of recurrence results to the general kernel K. These extensions will be discussed at the end of this chapter.

Throughout this chapter we assume that K is an irreducible kernel with maximal irreducibility measure ψ. R denotes the convergence parameter of K, $\mathscr{E}^+ = \{f \in \mathscr{E}_+ : \psi(f) > 0\}$, d is the period of K. By Theorem 2.1 there exist an integer $m_0 \geq 1$, a small function $s \in \mathscr{S}^+$ and small measure $\nu \in \mathscr{M}^+$ such that the minorization condition $M(m_0, 1, s, \nu)$ holds,

$$K^{m_0} \geq s \otimes \nu;$$

in other words, the pair (s, ν) is an atom for the kernel K^{m_0}. *We assume also that*

$$c_{m_0} = \text{g.c.d.}\{m_0, d\} = 1. \tag{5.1}$$

(By Remark 2.1(ii) this does not involve any loss of generality.) In this case K^{m_0} is an irreducible kernel on $(E, \mathscr{E})$ with period d and convergence parameter R^{m_0}. Moreover, K is R-recurrent if and only if K^{m_0} is R^{m_0}-recurrent (see Proposition 3.5).

5.1 Subinvariant and invariant functions

Let r, $0 < r < \infty$, be a constant.

Definition 5.1. A non-negative function $h \in \mathscr{E}^+$, which is not identically infinite, is called *r-subinvariant* (for K) if h is superharmonic for rK, i.e.,

$$h \geq rKh.$$

If $h \in \mathscr{E}^+, h \not\equiv \infty$, is harmonic for rK, i.e.,

$$h = rKh,$$

then h is called *r-invariant*.

We summarize some easy results on r-subinvariant functions in two propositions:

Proposition 5.1. Suppose that h is an r-subinvariant function. Then:

(i) The set $\{h < \infty\}$ is closed and full.
(ii) $h > 0$ everywhere.
(iii) If ν is a small measure then $0 < \nu(h) < \infty$.

Proof. (i) and (ii): Use Propositions 2.5 and 3.2.
(iii) By (ii), $\nu(h) > 0$. $\nu(h)$ is finite, since

$$h \geq r^{m_0+n} K^{m_0+n} h \geq r^{m_0+n} \nu(h) K^n s, \quad n \geq 0. \qquad \square$$

Proposition 5.2. (i) If either $r < R$, or $r = R$ and K is R-transient, then there exists an r-subinvariant function.
(ii) If $r > R$ then there does not exist any r-subinvariant function.

Proof. (i) Set $h = G^{(r)} s$ with s small.
(ii) Obvious. $\square$

The interesting, non-trivial case is the case where $r = R$ and K is R-recurrent. We denote by $G^{(R)}_{m_0,s,\nu}$ the potential kernel of $R^{m_0}(K^{m_0} - s \otimes \nu)$, i.e.,

$$G^{(R)}_{m_0,s,\nu} = \sum_{n=0}^{\infty} R^{nm_0}(K^{m_0} - s \otimes \nu)^n.$$

Note that, when $m_0 = 1$, i.e. (s, ν) is an atom, $G^{(R)}_{m_0,s,\nu}$ equals $G^{(R)}_{1,s,\nu} = G^{(R)}_{s,\nu}$ (cf. (4.13)).

Theorem 5.1. Suppose that K is R-recurrent. Then there exists an R-invariant function h_ν satisfying $\nu(h_\nu)=1$ and having the following uniqueness and minimality properties: For any R-subinvariant function h satisfying $\nu(h)=1$ we have

$$h=h_\nu \quad \psi\text{-a.e.,} \quad \text{and} \quad h \geq h_\nu \quad \text{everywhere.}$$

h_ν is given by the formula

$$h_\nu = R^{m_0} G^{(R)}_{m_0,s,\nu} s = \sum_{n=0}^{\infty} R^{(n+1)m_0}(K^{m_0} - s \otimes \nu)^n s.$$

Proof. Consider first the case where $m_0=1$. Then by Proposition 4.7(iii) $h_\nu \stackrel{\text{def}}{=} RG^{(R)}_{s,\nu} s$ satisfies $\nu(h_\nu)=\hat{b}(R)=1$. Consequently, $h_\nu \in \mathscr{E}^+$, h_ν is not identically infinite and

$$h_\nu = R(K - s\otimes\nu)h_\nu + Rs = RKh_\nu.$$

(We applied Proposition 2.1 to the kernel $R(K-s\otimes\nu)$.) Thus h_ν is R-invariant.

Let now $h \in \mathscr{E}^+$ be an arbitrary R-subinvariant function satisfying $\nu(h)=1$. We estimate h as follows:

$$\begin{aligned} h &\geq RKh = Rs + R(K - s\otimes\nu)h \\ &\geq h_\nu \quad \text{by Theorem 3.1(iii).} \end{aligned}$$

The function $h' = h - h_\nu \in \mathscr{E}_+$ satisfies $\nu(h')=0$, and $h' \geq RKh'$. By Proposition 5.1(ii), h' is not R-subinvariant. This is possible only if $h' \notin \mathscr{E}^+$, i.e. $h=h_\nu$ ψ-a.e.

Allow now m_0 to be arbitrary. What we have proved above holds true for the R^{m_0}-recurrent kernel K^{m_0}. Hence $h_\nu \stackrel{\text{def}}{=} R^{m_0} G^{(R)}_{m_0,s,\nu}$ is R^{m_0}-invariant for K^{m_0}, i.e.,

$$h_\nu = R^{m_0} K^{m_0} h_\nu.$$

It follows that

$$RKh_\nu = R^{m_0+1} K^{m_0+1} h_\nu.$$

Clearly $RKh_\nu > 0$ everywhere. Also, since $h_\nu = R^{m_0-1}K^{m_0-1}(RKh_\nu)$, RKh_ν cannot be identically infinite. Hence RKh_ν, too, is R^{m_0}-invariant for K^{m_0}. By minimality and uniqueness, there is a constant $c=\nu(RKh_\nu)$, $0<c<\infty$, such that

$$RKh_\nu \geq ch_\nu, \quad \text{with an equality } \psi\text{-a.e.}$$

By iterating we obtain

$$R^{m_0}K^{m_0}h_\nu \geq \cdots \geq c^{m_0-1}RKh_\nu \geq c^{m_0}h_\nu, \quad \text{with an equality } \psi\text{-a.e.}$$

Since h_ν is R^{m_0}-invariant for K^{m_0}, $c = 1$, and the above inequalities are equalities. In particular, h_ν is R-invariant for K.

The uniqueness and minimality of h_ν for K follow from the fact that every R-subinvariant function h for K is R^{m_0}-subinvariant for K^{m_0} and from the uniqueness and minimality of h_ν for K^{m_0}. □

For later purposes, note that when $m_0 = 1$ and $R = 1$, we have

$$h_\nu = \sum_{n=0}^{\infty} (K - s \otimes \nu)^n s = \sum_{n=0}^{\infty} a_n \quad \text{(see (4.11))}.$$

Specializing to the case where $K = P$ is the transition probability of a Markov chain (X_n) yields the following corollary. It generalizes Proposition 4.8 for arbitrary m_0. We write

$$G_{m_0,s,\nu} = G^{(1)}_{m_0,s,\nu} = \sum_{n=0}^{\infty} (P^{m_0} - s \otimes \nu)^n.$$

Corollary 5.1. Suppose that (X_n) is recurrent. Then the function $h_\nu = G_{m_0,s,\nu} s = \sum_{n=0}^{\infty} (P^{m_0} - s \otimes \nu)^n s$ is equal to the minimal harmonic function $\underline{h}$ introduced in Proposition 3.13. In particular, $\bar{H} = \{h_\nu = 1\}$ is the maximal Harris set for (X_n), and (X_n) is Harris recurrent if and only if $h_\nu \equiv 1$. □

Our next aim is to develop a useful tool, which enables us to transform almost any result from the probabilistic case $K = P$ to general K:

Let K be an arbitrary irreducible kernel with $R > 0$ and let h be an arbitrary r-subinvariant function (for some fixed $0 < r \leq R$). On the closed full set $F = \{h < \infty\} = \{0 < h < \infty\}$ (cf. Proposition 5.1) we can define a kernel $\tilde{K}$ by setting

$$\tilde{K}(x, A) = r(h(x))^{-1} \int_A K(x, dy) h(y), \quad x \in F, A \in \mathscr{E} \cap F;$$

or just briefly, denoting by I_h the (kernel of) *multiplication by h*,

$$\tilde{K} = r I_{h^{-1}} K I_h \quad \text{on } (F, \mathscr{E} \cap F). \tag{5.2}$$

We call (5.2) the *similarity transform of* K (by the r-subinvariant function h).

Clearly, $\tilde{K}$ is a transition probability on $(F, \mathscr{E} \cap F)$. Hence it governs the transitions of a Markov chain $(\tilde{X}_n; n \geq 0)$ with state space $(F, \mathscr{E} \cap F)$. The following result is obvious:

Proposition 5.3. (i) $(\tilde{X}_n)$ is irreducible with ψ as a maximal irreducibility measure.

(ii) The convergence parameter $\tilde{R}$ of $\tilde{K}$ is equal to $\tilde{R} = R/r$.

(iii) $(\tilde{X}_n)$ is recurrent if and only if $r = R$ and K is R-recurrent. □

The most interesting special case is the case where $r = R$, K is R-recurrent and $h = h_\nu$ is the minimal R-invariant function given by Theorem 5.1.

Proposition 5.4. Suppose that K is R-recurrent. Let $h = h_\nu = \sum_{n=0}^{\infty} R^{(n+1)m_0}(K^{m_0} - s\otimes \nu)^n s$, $\tilde{E} = \{h_\nu < \infty\}$, $\tilde{K} = RI_{(h_\nu)^{-1}} K I_{h_\nu}$ on $\tilde{E}$. Then the Markov chain $(\tilde{X}_n)$ is Harris recurrent (on its state space $\tilde{E}$).

Proof. There is no loss of generality in assuming that $h_\nu < \infty$ everywhere. (Otherwise restrict K to $\tilde{E}$.)

By Proposition 3.13 we have to prove that the minimal superharmonic function $\underline{\tilde{h}}$ for $(\tilde{X}_n)$ is identically equal to 1.

Since h_ν is R-invariant for K, 1 is harmonic for $(\tilde{X}_n)$. Let now $\tilde{h}$ be superharmonic for $(\tilde{X}_n)$ satisfying $\tilde{h} = 1$ ψ-a.e. Then $h_\nu \tilde{h}$ is R-subinvariant for K, and, by the minimality of h_ν, we have $h_\nu \tilde{h} \geq h_\nu$. Hence $\tilde{h} \geq 1$, and therefore $\underline{\tilde{h}} \equiv 1$. □

For later purposes we note the following

Proposition 5.5. Suppose that K is R-recurrent. Then for any m, $i \geq 1$, h_ν restricted to the set $E_i^{(m)} = E_i + E_{i+c_m} + \cdots + E_{i+d-c_m}$ is the (essentially) unique minimal R^m-invariant function for the R^m-recurrent kernel K^m with state space $E_i^{(m)}$.

Proof. Use similar arguments as in the proof of Proposition 3.14. □

5.2 Subinvariant and invariant measures

Let $0 < r < \infty$ be a constant.

Definition 5.2. A measure $\pi \in \mathcal{M}^+$, $\pi(B) < \infty$ for some $B \in \mathcal{E}^+$, is called *r-subinvariant* (for K), if

$$\pi \geq r\pi K.$$

If

$$\pi = r\pi K,$$

then the r-subinvariant measure π is called *r-invariant.*

When $r = 1$ we say simply *subinvariant* (resp. *invariant*) instead of 1-subinvariant (resp. 1-invariant).

Proposition 5.6. Suppose that π is an r-subinvariant measure. Then:

(i) π is σ-finite and $\psi \ll \pi$.
(ii) If s is a small function then $0 < \pi(s) < \infty$.

Proof. Let $B \in \mathcal{E}$ be arbitrary. Then

$$\pi(B) \geq r^n \pi K^n(B) \quad \text{for all } n \geq 0. \tag{5.3}$$

Let $f = G^{(\rho)}1_B$ for some $0 < \rho < r$. If $B \in \mathscr{E}^+$ is such that $\pi(B) < \infty$ (cf. Definition 5.2) then it follows that $\pi(f) < \infty$. By irreducibility $f > 0$ everywhere. Hence π is σ-finite.

By (5.3) and by irreducibility, $\pi(B) > 0$ for all $B \in \mathscr{E}^+$. Thus ψ is absolutely continuous w.r.t. π.

The proof of (ii) is similar to that of Proposition 5.1(iii). □

Analogously to Proposition 5.2 we have:

Proposition 5.7. (i) If either $r < R$, or $r = R$ and K is R-transient, then there exists an r-subinvariant measure which is equivalent to ψ (i.e., serves as a maximal irreducibility measure).

(ii) If $r > R$ then there does not exist any r-subinvariant measure.

Proof. (i) Set $\pi = \nu G^{(r)}$, ν any small measure.

(ii) Obvious. □

We omit the proof of the following theorem since it is completely analogous to that of Theorem 5.1.

Theorem 5.2. Suppose that K is R-recurrent. Then the measure π_s defined by

$$\pi_s = R^{m_0}\nu G^{(R)}_{m_0,s,\nu} = \sum_{n=0}^{\infty} R^{(n+1)m_0}\nu(K^{m_0} - s \otimes \nu)^n$$

is R-invariant, is equivalent to ψ, and satisfies $\pi_s(s) = 1$. It is the unique R-subinvariant measure π satisfying $\pi(s) = 1$. □

Suppose that K is R-recurrent. In what follows we shall use the symbol π to denote an R-invariant measure for K. By the above theorem there is a constant $c = \pi(s)$, $0 < c < \infty$, such that

$$\pi = c\pi_s, \tag{5.4}$$

and π is a maximal irreducibility measure. So, for example,

$$\mathscr{E}^+ = \{A \in \mathscr{E} : \pi(A) > 0\},$$

and a set $A \in \mathscr{E}$ is full if and only if $\pi(A^c) = 0$.

Let h be an arbitrary R-subinvariant function for K. Since by Theorem 5.1 there is a constant $c = \nu(h)$, $0 < c < \infty$, such that

$$h = ch_\nu \quad \pi\text{-a.e.}, \tag{5.5}$$

the following definition is unambiguous:

Definition 5.3. An R-recurrent kernel K is called *R-positive recurrent* if $\pi(h)$ is finite, otherwise *R-null recurrent.*

By (5.4) and (5.5), and by the definitions of π_s and h_ν, the integral $\pi(h)$ takes the form

$$\pi(h) = \pi(s)\nu(h) \sum_{n=0}^{\infty} (n+1)R^{(n+2)m_0}\nu(K^{m_0} - s \otimes \nu)^n s. \tag{5.6}$$

In the case where $m_0 = 1, \pi(h)$ can also be written in the form

$$\pi(h) = R^2\pi(s)\nu(h)\frac{\mathrm{d}\hat{b}(R)}{\mathrm{d}R},$$

whenever $\hat{b}$ is differentiable at R. In fact, it is easy to see that the left hand side derivative

$$\lim_{r\uparrow R}\uparrow \frac{1-\hat{b}(r)}{R-r} \text{ always exists and equals}$$

$$R^{-2}\pi(s)^{-1}\nu(h)^{-1}\pi(h) \quad (\leq \infty).$$

Suppose now that $K = P$ is the transition probability of a Markov chain (X_n). As a corollary to Theorem 5.2 we obtain:

Corollary 5.2. Suppose that (X_n) is recurrent. Then the measure π_s defined by

$$\pi_s = \nu G_{m_0,s,\nu} = \sum_{n=0}^{\infty} \nu(P^{m_0} - s \otimes \nu)^n$$

is invariant, is equivalent to ψ, and satisfies $\pi_s(s) = 1$. Moreover, it is the unique subinvariant measure π satisfying $\pi(s) = 1$. □

We call the Markov chain (X_n) *positive recurrent* or *null recurrent*, depending on whether the transition probability P is 1-positive recurrent or 1-null recurrent. Note that, since the harmonic functions for a recurrent Markov chain are constants (π-a.e.), positive recurrence is equivalent to π being a finite measure. In this case we always norm π to a probability measure, and call it also the *stationary distribution* of the Markov chain (X_n).

It is easy to see that in the positive recurrent case the Markov chain $(X_n; n \geq 0)$ is a stationary process, given that the initial distribution is the stationary distribution, $\mathscr{L}(X_0) = \pi$.

From (5.6) we get

$$\pi(E) = \pi(s) \sum_{n=0}^{\infty} (n+1)\nu(P^{m_0} - s \otimes \nu)^n s.$$

When $m_0 = 1$, the invariant measure can be interpreted in terms of the split chain (X_n, Y_n):

$$\pi_s(A) = \sum_0^\infty \nu(P - s \otimes \nu)^n(A) = \sum_0^\infty \sigma_n(A)$$
$$= \mathbb{E}_\alpha \sum_1^{S_\alpha} 1_A(X_n), \quad A \in \mathscr{E}. \tag{5.7}$$

Hence $\pi_s(A)$ is the expected number of visits by (X_n) to the set A during an α-*block*, that is, between two consecutive occurrences of the event $\{Y_n = 1\}$. In particular, we obtain

$$\pi_s(E) = \mathbb{E}_\alpha S_\alpha.$$

Thus (X_n) is positive recurrent if and only if the expectation $\mathbb{E}_\alpha S_\alpha$ is finite.

Let K be an R-recurrent kernel with R-invariant function h and measure π, and let $\tilde{K}$ denote the transformed kernel $\tilde{K} = RI_{h^{-1}}KI_h$. Since clearly the measure $\tilde{\pi} = \pi I_h$ is invariant for $\tilde{K}$, we obtain the following result:

Proposition 5.8. The kernel K is R-positive recurrent if and only if the Markov chain $(\tilde{X}_n)$ with transition probability $\tilde{K}$ is positive recurrent. □

Examples 5.1. (*a*) Let (X_n) be a recurrent, discrete Markov chain with transition matrix P and with maximal irreducibility measure $\psi = \text{Card}_F$. Then, for any fixed state $z \in F$, the row vector π_z defined by

$$\pi_z(x) = \mathbb{E}_z \sum_1^{S_z} 1_{\{X_n = x\}}, \quad x \in E,$$

is invariant. Moreover, (X_n) is positive recurrent if and only if the expectation $\mathbb{E}_x S_x$ is finite for some $x \in E$. Then $\mathbb{E}_x S_x$ is finite whenever $x \in F$; the stationary distribution π is given by

$$\pi(x) = (\mathbb{E}_x S_x)^{-1}, \quad x \in E.$$

Henceforth, when dealing with a recurrent, discrete Markov chain, we shall write E_π for the set of π-positive states, $E_\pi = \{x \in E : \pi(x) > 0\}$ $(= F)$.

(*c*) The Lebesgue measure ℓ is an invariant measure for the random walk on $(\mathbb{R}, \mathscr{R})$.

(*e*) The measure π defined by $\pi(\mathrm{d}t) = (1 - F(t))\mathrm{d}t$ is invariant for the Markov chain $(V_{n\delta}^+; n \geq 0)$. Hence $(V_{n\delta}^+)$ (with spread-out F) is positive recurrent if and only if $\mathbb{E}z_1 = \int_{\mathbb{R}_+} (1 - F(t))\mathrm{d}t$ is finite.

5.3 Expectations over blocks

Throughout Sections 5.3–5.5 we assume that $K = P$ is the transition probability of a Harris recurrent Markov chain (X_n). π denotes a fixed invariant measure for (X_n). (There is not much loss of generality in assuming the somewhat stronger Harris recurrence instead of mere recurrence. If (X_n) were only recurrent we should consider its restriction to a Harris set $H(=E$ π-a.e.).)

Note, in particular, that P is stochastic, and therefore

$$\mathbb{E}_x S_B = U_B(x, E) \quad \text{for all } x \in E, B \in \mathscr{E}.$$

For any $B \in \mathscr{E}$, let $0 \leq T_B(0) < T_B(1) < \cdots$ denote the successive visit epochs by (X_n) to B. For any $i \geq 1$, the sequence $(X_{T_B(i-1)+1}, \ldots, X_{T_B(i)})$ is called the *i'th B-block*. By the Markov property, for any $i \geq 1$ and $x \in B$, given the history up to $T_B(i-1)$ and given $X_{T_B(i-1)} = x$, the ith B-block has the same distribution as the first B-block $(X_1, \ldots, X_{S_B})$ given $X_0 = x$. In particular, it follows that for any $f \in \mathscr{E}_+$:

$$\mathbb{E}\left[\sum_{n=T_B(i-1)+1}^{T_B(i)} f(X_n) \,|\, \mathscr{F}_m^X; T_B(i-1) = m, X_m = x\right] = \mathbb{E}_x \sum_1^{S_B} f(X_n)$$
$$= U_B f(x), \quad \text{for all } x \in B, m \geq 0, i \geq 1.$$

When P has an atom (s, ν) we can construct the split chain (X_n, Y_n) in the manner described in Section 4.4. By the ith α-block we mean the sequence $(X_{T_\alpha(i-1)+1}, \ldots, X_{T_\alpha(i)})$. The α-blocks are i.i.d. random elements having the same distribution as the sequence $(X_1, \ldots, X_{S_\alpha})$ given $Y_0 = 1$. In particular, for any $f \in \mathscr{E}_+$:

$$\mathbb{E}\left[\sum_{n=T_\alpha(i-1)+1}^{T_\alpha(i)} f(X_n) \,|\, \mathscr{F}_{m-1}^{X,Y}; T_\alpha(i-1) = m\right] = \mathbb{E}_\alpha \sum_1^{S_\alpha} f(X_n)$$
$$= \pi_s(f) \quad \text{by (5.7)}$$
$$= \pi(s)^{-1} \pi(f),$$

i.e., the expectation of $f(X_n)$ summed over any α-block is equal (up to multiplication by a constant) to the invariant measure $\pi(f)$ of f.

We shall show that a similar result holds true for a B-block, provided that the Markov chain (X_n) starts 'stationarily' from B.

Recall from (3.6) the definition of the potential kernel

$$G'_B(x, A) = \sum_0^\infty (P I_{B^c})^n (x, A) = \mathbb{E}_x \sum_{n=0}^{S_B - 1} 1_A(X_n).$$

Proposition 5.9. For all $B \in \mathscr{E}^+ : \pi I_B G'_B = \pi I_B U_B = \pi$.

Proof. Denote

$$\pi' = \pi I_B G'_B.$$

Since $\pi I_B U_B = \pi' P$, we need only show that $\pi' = \pi$.

Decompose π as follows:

$$\begin{aligned}\pi &= \pi I_B + \pi P I_{B^c}, \quad \text{since } \pi = \pi P,\\ &= \pi I_B \sum_0^{N-1} (P I_{B^c})^n + \pi (P I_{B^c})^N, \quad \text{by iterating},\\ &\geq \pi', \quad \text{letting } N \to \infty.\end{aligned}$$

Further, π' is subinvariant:

$$\begin{aligned}\pi' &= \pi I_B + \pi' P I_{B^c} \quad \text{by the definition of } \pi'.\\ &= \pi P I_B + \pi' P I_{B^c} \quad \text{since } \pi = \pi P,\\ &\geq \pi' P \quad \text{since } \pi \geq \pi'.\end{aligned}$$

By the definition of π', the subinvariant measures π and π' coincide on $(B, \mathscr{E} \cap B)$. It follows from the uniqueness of π (see Theorem 5.2) that $\pi' = \pi$. □

The result of Proposition 5.9 can also be written in the form

$$\begin{aligned}\int_B \pi(\mathrm{d}x) \mathbb{E}_x \sum_0^{S_B - 1} f(X_n) &= \int_B \pi(\mathrm{d}x) \mathbb{E}_x \sum_1^{S_B} f(X_n)\\ &= \pi(f), \quad \text{for all } f \in \mathscr{E}_+.\end{aligned}$$

Setting $f \equiv 1$, we obtain the following corollary. For its part (ii), recall from Proposition 5.6 that $0 < \pi(C) < \infty$ for every small set $C \in \mathscr{S}^+$.

Corollary 5.3. (i) For all $B \in \mathscr{E}^+$:

$$\int_B \pi(\mathrm{d}x) \mathbb{E}_x S_B = \pi(E) < \infty \quad \text{or } = \infty$$

depending on whether (X_n) is positive or null recurrent.

(ii) If

$$\sup_{x \in C} \mathbb{E}_x S_C < \infty \quad \text{for some small set } C \in \mathscr{S}^+,$$

then (X_n) is positive recurrent. □

The following proposition gives a criterion for positive recurrence in terms of a 'drift condition':

Proposition 5.10. Let $B \in \mathscr{E}$ be arbitrary. Suppose that for some function $g \in \mathscr{E}_+$ and constant $\gamma > 0$:

$$\begin{aligned}&\mathbb{E}[g(X_{n+1}) - g(X_n) \mid X_n = x]\\ &\qquad = Pg(x) - g(x) \leq -\gamma \quad \text{for all } x \in B^c. \qquad (5.8)\end{aligned}$$

Then

$$\mathbb{E}_x S_B \le \gamma^{-1} P I_{B^c} g(x) + 1 \quad \text{for all } x \in E$$

and

$$\mathbb{E}_x S_B \le \gamma^{-1} g(x) \quad \text{for all } x \in B^c.$$

In particular, if $C \in \mathscr{E}^+$ is small, and

$$\sup_{x \in C^c} \mathbb{E}[g(X_{n+1}) - g(X_n) | X_n = x] < 0,$$

and

$$\sup_{x \in C} \mathbb{E}[g(X_{n+1}); X_{n+1} \in C^c | X_n = x] = \sup_{x \in C} P I_{C^c} g(x) < \infty,$$

then (X_n) is positive recurrent.

Proof. The hypothesis (5.8) can be written in the form

$$\begin{aligned} g &\ge I_{B^c} P g + \gamma 1_{B^c} \\ &\ge \gamma G_B 1_{B^c} \quad \text{by Theorem 3.1(iii)}, \\ &= \gamma 1_{B^c}(x) \mathbb{E}_x S_B. \end{aligned}$$

'Multiplying' this from the left by $P I_{B^c}$ we get

$$P I_{B^c} g(x) \ge \gamma U_B 1_{B^c}(x) = \gamma(\mathbb{E}_x S_B - 1)$$

The rest follows from Corollary 5.3(ii). □

Example 5.2. (*d*) (The reflected random walk). Suppose that $\mathbb{E} z_1 < 0$. Then (W_n) is positive recurrent. (Hint: Choose $g(x) = x$, and $C = [0, c]$, c sufficiently large, in Proposition 5.10.)

Our next object of study is the sums over the sequence $(X_0, \ldots, X_{S_B})$. The following concepts and results will be used later in Chapters 6 and 7 in the investigation of the asymptotic behaviour of the n-step transition probabilities P^n.

We write $\mathscr{L}^1(\pi)$ for the set of π-integrable functions $f \in \mathscr{E}$.

Let $B \in \mathscr{E}^+$ and $f \in \mathscr{L}^1_+(\pi)$ be arbitrary. It follows from Proposition 5.9 that $U_B f(x) = \mathbb{E}_x \sum_1^{S_B} f(X_n)$ is finite for π-almost all $x \in B$. According to the following proposition $U_B f$ is finite even on a full set (i.e. π-a.e.).

Proposition 5.11. For any $B \in \mathscr{E}^+, f \in \mathscr{L}^1_+(\pi)$: $U_B f$ is finite π-almost everywhere.

Proof. Since $U_B f$ is finite π-a.e. on B, and

$$G_B f = f + I_{B^c} U_B f,$$

it is sufficient to show that $G_B f$ is finite π-a.e. By Proposition 2.5(ii) there is a closed set F such that $U_B f$ is finite on $B \cap F$. We claim that the set $F' = \{G_B f < \infty\} \cap F$ is closed.

If $x \in B \cap F' \subseteq B \cap F$ then

$$\infty > U_B f(x) = PG_B f(x),$$

whence $P(x, (F')^c) = 0$. If $x \in B^c \cap F'$ then

$$\infty > G_B f(x) \geq PG_B f(x),$$

whence again $P(x, (F')^c) = 0$. It follows that F' is closed. By Proposition 2.5(i) it is full. □

Setting $f \equiv 1$ we obtain the following:

Corollary 5.4. If (X_n) is positive recurrent then for all $B \in \mathscr{E}^+$, $\mathbb{E}_x S_B$ is finite for π-almost all $x \in E$. □

Let us fix a non-negative π-integrable function $f \in \mathscr{L}^1_+(\pi)$. The full set $\{U_B f < \infty\}$ may of course depend on B. We set the following:

Definition 5.4. A *state* $x \in E$ is called *f-regular* if $f(x)$ and $U_B f(x)$ are finite for all $B \in \mathscr{E}^+$.

More generally, a *finite measure* $\lambda \in b\mathscr{M}^+$ is called *f-regular* if $\lambda(f)$ and $\lambda U_B f$ are finite for all $B \in \mathscr{E}^+$. A *subset* $D \subseteq E$ is called *f-regular* if f and $U_B f$ are bounded on D for all $B \in \mathscr{E}^+$.

A bounded π-integrable function $g \in b\mathscr{L}^1(\pi)$ is called *special* if the whole state space E is $|g|$-regular, i.e., if $U_B|g|$ is bounded for all $B \in \mathscr{E}^+$. A set $D \in \mathscr{E}$ with $\pi(D) < \infty$ is called special if 1_D is a special function.

When $f \equiv 1$ we say simply 'regular' instead of '1-regular'.

We denote by R_f the set of f-regular states $x \in E$.

It is by no means trivial that there should exist any f-regular states. However, we shall show that there exists even an increasing sequence $D_1 \subseteq D_2 \subseteq \cdots$ of f-regular sets such that their union $\bigcup_1^\infty D_i$ is full.

In order to avoid unessential technicalities *we assume for the rest of this section that* (X_n) *is aperiodic*. Recall the definition of the kernel

$$G_{m_0,s,\nu} = \sum_{n=0}^{\infty} (P^{m_0} - s \otimes \nu)^n,$$

and the probabilistic interpretation of the kernel $G_{s,\nu} = G_{1,s,\nu}$:

$$G_{s,\nu} f(x) = \mathbb{E}_x \sum_{n=0}^{T_\alpha} f(X_n).$$

Let us denote

$$\bar{G}_{m_0,s,v} = G_{m_0,s,v}(I + P + \cdots + P^{m_0 - 1})$$
$$= \sum_{n=0}^{\infty} (P^{m_0} - s \otimes v)^n (I + P + \cdots + P^{m_0 - 1}),$$

noting that $\bar{G}_{1,s,v} = G_{s,v}$.

Proposition 5.12. $\bar{G}_{m_0,s,v}f$ is finite π-almost everywhere.

Proof. If $m_0 = 1$ then the result follows from Proposition 5.11 by applying it to the split chain (X_n, Y_n) with $B = \alpha = E \times \{1\}$.

By considering the m_0-step chain (X_{nm_0}) and the function $(I + P + \cdots + P^{m_0 - 1})f$ instead of f we can conclude that the result holds true also in the case $m_0 \geq 2$. □

Proposition 5.13. (i) A set $D \subseteq E$ is f-regular if and only if $\bar{G}_{m_0,s,v}f$ is bounded on D.

(ii) In particular, the set of f-regular states R_f is equal to the full set $\{\bar{G}_{m_0,s,v}f < \infty\}$. There exists an increasing sequence $D_1 \subseteq D_2 \subseteq \cdots$ of f-regular sets, $D_i = \{\bar{G}_{m_0,s,v}f \leq i\}$, $i \geq 1$, for example, such that $\bigcup_1^\infty D_i = R_f$.

(iii) A function $g \in b\mathscr{L}^1(\pi)$ is special if and only if $\bar{G}_{m_0,s,v}|g|$ is bounded. In particular, every small function s is special, and there is an increasing sequence $D_1 \subseteq D_2 \subseteq \ldots$ of special sets such that $\bigcup_1^\infty D_i = E$.

(iv) A finite measure λ is f-regular if and only if $\lambda\bar{G}_{m_0,s,v}f$ is finite.

Proof. Suppose first that $m_0 = 1$, i.e., (s, v) is an atom. Since $S_B \leq T_\alpha + S_B \circ \theta_{T_\alpha}$ we obtain.

$$f(x) + U_B f(x) = \mathbb{E}_x \sum_0^{S_B} f(X_n)$$
$$\leq \mathbb{E}_x \sum_0^{T_\alpha} f(X_n) + \mathbb{E}_\alpha \sum_1^{S_B} f(X_n).$$

The first term on the right hand side equals $G_{s,v}f(x)$. By applying Proposition 5.11 to the split chain, we see that the second term is finite. Hence, if $\bar{G}_{1,s,v}f = G_{s,v}f$ is bounded on D then f and $U_B f$ are bounded on D; i.e., D is f-regular.

Conversely, if D is f-regular, we take a set $D' \in \mathscr{E}^+$ such that $G_{s,v}f$ is bounded on D' (cf. Proposition 5.12) and use the inequality $T_\alpha \leq S_{D'} + T_\alpha \circ \theta_{S_{D'}}$ to prove that $G_{s,v}$ is bounded on D. Thus the proof of (i) is completed in the case where $m_0 = 1$. Note that (ii) and (iii) are immediate consequences of (i). (For (iii) recall also Proposition 2.11(iv).) The proof of (iv) is similar to that of (i).

In order to prove the case $m_0 \geq 2$ we need three lemmas:

The first lemma gives a simple criterion in terms of 'geometric trials' for the finiteness of a stopping time.

Lemma 5.1. Let $(\mathscr{F}_n; n \geq 0)$ be a history and let τ be a stopping time relative to it. Suppose that there is a constant $\gamma > 0$ such that

$$\mathbb{P}\{\tau = n | \mathscr{F}_{n-1}\} \geq \gamma \quad \text{on } \{\tau \geq n\}, \quad \text{for all } n \geq 1.$$

Then τ is a finite (a.s.) and

$$\mathbb{E}[\tau | \mathscr{F}_0] \leq \gamma^{-1} < \infty.$$

Proof. By the hypotheses

$$\begin{aligned}\mathbb{P}\{\tau \geq n+1 | \mathscr{F}_0\} &= \mathbb{E}[\mathbb{P}\{\tau \neq n | \mathscr{F}_{n-1}\}; \tau \geq n | \mathscr{F}_0] \\ &\leq (1-\gamma)\mathbb{P}\{\tau \geq n | \mathscr{F}_0\} \quad \text{for all } n \geq 1.\end{aligned}$$

Consequently,

$$\mathbb{P}\{\tau \geq n | \mathscr{F}_0\} \leq (1-\gamma)^{n-1} \quad \text{for all } n \geq 1,$$

from which the results follow. □

The second lemma can be regarded as a generalization of the classical Wald lemma. Wald's lemma states that, if $\xi_0, \xi_1, \ldots$ are i.i.d. random variables with common finite mean M, and τ is a stopping time relative to the internal history $(\mathscr{F}_n^{\xi})$ and with finite mean $\mathbb{E}\tau$, then the expectation

$$\mathbb{E}\sum_0^{\tau} \xi_n = M(1 + \mathbb{E}\tau)$$

is finite.

Lemma 5.2. Suppose that $(\xi_n; n \geq 0)$ is a non-negative stochastic process, adapted to a history $(\mathscr{F}_n; n \geq 0)$. Let τ be a stopping time relative to $(\mathscr{F}_n)$ and let

$$M_0 = \mathbb{E}\xi_0, \quad M = \sup_{n \geq 1}\{\text{ess sup } \mathbb{E}[\xi_n | \mathscr{F}_{n-1}]\}.$$

Then

$$\begin{aligned}\mathbb{E}\sum_0^{\tau} \xi_n &= \mathbb{E}\xi_0 + \mathbb{E}\left[\sum_1^{\tau} \mathbb{E}[\xi_n | \mathscr{F}_{n-1}]; \tau \geq 1\right] \\ &\leq M_0 + M\mathbb{E}\tau.\end{aligned}$$

Proof. Write

$$\mathbb{E}\sum_0^{\tau} \xi_n = \mathbb{E}\xi_0 + \sum_1^{\infty} \mathbb{E}[\xi_n; \tau \geq n]$$

and condition w.r.t. the σ-algebras $\mathscr{F}_{n-1}$. □

The last lemma also completes the proof of Proposition 5.13:

Lemma 5.3. A set $D \subseteq E$ (or a measure λ) is f-regular if and only if it is $(I + \cdots + P^{m_0 - 1})f$-regular for the m_0-step chain $(X_{nm_0}; n \geq 0)$.

Proof. Set $m = m_0$ and define

$${}_mS_B = \inf\{n \geq 1 : X_{nm} \in B\}.$$

By using the Markov property at nm we easily get the following identity: for any $x \in E$, $B \in \mathscr{E}^+$,

$$\mathbb{E}_x \sum_{n=0}^{{}_mS_B} (f + \cdots + P^{m-1} f)(X_{nm}) = \mathbb{E}_x \sum_{n=0}^{m_mS_B + m - 1} f(X_n). \tag{5.9}$$

If $D \subseteq E$ is $(f + \cdots + P^{m-1} f)$-regular for (X_{nm}), it follows from (5.9) and the inequality $S_B \leq m_mS_B$ that D is f-regular for (X_n).

To prove the converse, suppose that $D \subseteq E$ is f-regular. Proposition 5.13 applied to the m-step chain states that there exists a set $C \in \mathscr{E}^+$ which is $(f + \cdots + P^{m-1} f)$-regular for (X_{nm}). (Note that the m-step chain possesses the atom (s, ν) and we have already proved Proposition 5.13 for Markov chains having an atom.) By the first part of the present proof, C is f-regular for (X_n). By Proposition 2.6 we may suppose that C is also a small set. This and aperiodicity imply that there is an integer $n_0 \geq 2m$ and constant $\gamma > 0$ such that (keeping $B \in \mathscr{E}^+$ fixed)

$$P^n(y, B) \geq \gamma \quad \text{for all } y \in C,\ n_0 - 2m < n \leq n_0 - m. \tag{5.10}$$

Define a sequence $(\eta(i); i \geq 0)$ of stopping times by setting $\eta(0) = T_C$, and

$$\eta(i) = \inf\{n \geq \eta(i-1) + n_0 : X_n \in C\}, \quad i \geq 1.$$

Since (X_n) is Harris recurrent by our basic hypothesis, $\eta(i)$ is finite $\mathbb{P}_x$-a.s. for all $x \in E$, $i \geq 1$.

For every $i \geq 1$, set $(\mathscr{F}_i) = (\mathscr{F}^X_{\eta(i)})$, and

$$\xi_0 = \sum_0^{\eta(0)} f(X_n) = \sum_0^{T_C} f(X_n), \quad \xi_i = \sum_{\eta(i-1)+1}^{\eta(i)} f(X_n).$$

Let $\sigma(i)$ be the unique integer satisfying

$$\eta(i-1) + n_0 - 2m < \sigma(i)m \leq \eta(i-1) + n_0 - m.$$

Define a stopping time τ relative to the history $(\mathscr{F}_i)$ by

$$\tau = \inf\{i \geq 1 : X_{\sigma(i)m} \in B\}.$$

It follows from (5.10) that

$$\mathbb{P}\{\tau = i \mid \mathscr{F}_{i-1}\} \geq \gamma \quad \text{on } \{\tau \geq i\}, \quad \text{for all } i \geq 1.$$

Since D is f-regular,

$$M_0 \overset{\text{def}}{=} \sup_{x \in D} \mathbb{E}_x \xi_0 = \sup_{x \in D} G_C f(x) < \infty.$$

Applying the strong Markov property at $\eta(i-1)$, we obtain for all $x \in E$, $i \geq 1$,

$$\mathbb{E}_x[\xi_i | \mathscr{F}_{i-1}] \leq \sup_{y \in D} \mathbb{E}_y \sum_1^{S'_C} f(X_n),$$

where $S'_C = \inf\{n \geq n_0 : X_n \in C\} \leq S_C(n_0)$.

Consequently, by using the strong Markov property $n_0 - 1$ times,

$$\mathbb{E}_x[\xi_i | \mathscr{F}_{i-1}] \leq \sup_{y \in C} \sum_1^{S_C(n_0)} f(X_n) \leq n_0 \sup_{y \in C} U_C f(y) = M, \text{ say.}$$

M is finite, because C is f-regular. By applying Lemmas 5.1 and 5.2 and using the inequality $m_m S_B + m \leq \eta(\tau)$, we obtain

$$\sup_{x \in D} \mathbb{E}_x \sum_{n=0}^{m_m S_B + m - 1} f(X_n) \leq \sup_{x \in D} \mathbb{E}_x \sum_{i=0}^{\tau} \xi_i \leq M_0 + M\gamma^{-1} < \infty,$$

That D is $(f + \cdots + P^{m-1} f)$-regular for (X_{nm}) now follows from the identity (5.9).

The result for λ is proved similarly. □

Let $A \in \mathscr{E}^+$ be an f-regular set, and let $D \subseteq E$ be arbitrary. By using the inequality $S_B \leq T_A + S_B \circ \theta_{T_A}$ we see that, if $G_A f(x) = \mathbb{E}_x \sum_0^{T_A} f(X_n)$ is bounded on D then also D is f-regular. Since by Proposition 5.13 any set $B_0 \in \mathscr{E}^+$ contains an f-regular set we obtain the following criterion for f-regularity:

Proposition 5.14. (i) A set $D \subseteq E$ is f-regular if (and only if) for some f-regular set $A \in \mathscr{E}^+$:

$$\sup_D G_A f < \infty. \tag{5.11}$$

(ii) A set $D \subseteq E$ is f-regular, if (and only if) for some set $B_0 \in \mathscr{E}^+$, (5.11) holds for all $A \in \mathscr{E}^+ \cap B_0$. □

A similar result holds true for measures. For special functions we have:

Corollary 5.5. (i) A π-integrable function $g \in \mathscr{L}^1(\pi)$ is special if (and only if) for some $|g|$-regular set $A \in \mathscr{E}^+$:

$$\sup_E G_A |g| < \infty. \tag{5.12}$$

(ii) A π-integrable function $g \in \mathscr{L}^1(\pi)$ is special if (and only if) for some set $B_0 \in \mathscr{E}^+$, (5.12) holds for all $A \in \mathscr{E}^+ \cap B_0$.

Examples 5.3. (*a*) Let (X_n) be a positive Harris recurrent, discrete Markov chain with stationary distribution π (cf. Example 5.1(*a*)). The set R_1 of regular states is equal to $R_1 = \{x\in E\colon \mathbb{E}_x S_{x_0} < \infty\}$ ($\supseteq E_\pi$), where $x_0 \in E_\pi$ is arbitrary.

(d) For a positive recurrent reflected random walk (see Example 5.2 (d)) every state $x\in\mathbb{R}_+$ is regular. More generally, every bounded set $D\subset\mathbb{R}_+$ is regular.

(e) (The forward process). Suppose that the Markov chain $(V^+_{n\delta}; n\geq 0)$ is positive recurrent (cf. Example 5.1(e)). Then every bounded set $D\subset\mathbb{R}_+$ is regular. An initial distribution F_0 is regular if and only if the mean $\int tF_0(\mathrm{d}t)$ is finite.

5.4 Recurrence of degree 2

Let (X_n) be a Harris recurrent Markov chain. It follows from Corollary 5.3(i) that positive recurrent chains can be characterized as those recurrent chains for which some (or equivalently, all) of the expectations $\int_B \pi(\mathrm{d}x)\mathbb{E}_x S_B$, $B\in\mathscr{E}^+$, is (are) finite.

Consideration of the second moments of the hitting times S_B leads to the following:

Definition 5.5. The Markov chain (X_n) is called *recurrent of degree* 2, if

$$\int_B \pi(\mathrm{d}x)\mathbb{E}_x S_B^2 < \infty \quad \text{for all } B\in\mathscr{E}^+.$$

Note that, by Corollary 5.3(i) recurrence of degree 2 implies positive recurrence. Below we will see that recurrence of degree 2 can be alternatively characterized as the regularity of the stationary distribution π. In order to prove this we need to consider the following identity:

Lemma 5.4. For all $B\in\mathscr{E}$:

$$\mathbb{E}_x[\tfrac{1}{2}S_B(S_B+1)] = U_B G_B 1(x) \quad \text{for all } x\in E. \tag{5.13}$$

Proof. Since

$$\tfrac{1}{2}S_B(S_B+1) = \sum_0^{S_B}(S_B-n),$$

and $S_B = n + T_B\circ\theta_n$ on $\{S_B\geq n\}$ for all $n\geq 1$, the left hand side of (5.13) is equal to

$$\begin{aligned}\mathbb{E}_x S_B &+ \mathbb{E}_x \sum_{n=1}^{S_B} T_B\circ\theta_n\\ &= \mathbb{E}_x S_B + \mathbb{E}_x \sum_{n=1}^{S_B} \mathbb{E}_{X_n} T_B, \text{ by conditioning w.r.t. } \mathscr{F}_n,\\ &= U_B 1(x) + U_B(G_B 1 - 1)(x) = U_B G_B 1(x). \qquad \square\end{aligned}$$

Proposition 5.15. For all $B \in \mathscr{E}^+$:

$$\mathbb{E}_{\pi I_B}[S_B^2] = 2\mathbb{E}_\pi T_B + 1 = 2\mathbb{E}_\pi S_B - 1.$$

Proof. Integrate both sides of (5.13) over B w.r.t. the measure π, and use Proposition 5.9. $\square$

For the rest of this section we assume that (X_n) is aperiodic.

By using Propositions 5.13, 5.14 and 5.15 we get the following characterizations for recurrence of degree 2:

Proposition 5.16. Each of the following four conditions is equivalent to recurrence of degree 2:

(i) The invariant probability measure π is regular, i.e., $\mathbb{E}_\pi S_B$ is finite for all $B \in \mathscr{E}^+$.

(ii) $\pi \bar{G}_{m_0,s,v}(E)$ is finite.

(iii) $\mathbb{E}_{\pi I_D}[S_D^2]$ is finite for some regular set $D \in \mathscr{E}^+$.

(iv) There is a set $B_0 \in \mathscr{E}^+$ such that $\mathbb{E}_{\pi I_A}[S_A^2]$ is finite for all $A \in \mathscr{E}^+ \cap B_0$. $\square$

For later purposes we introduce the concept of regularity of degree 2:

Definition 5.6. A state $x \in E$ (resp. a measure $\lambda \in \mathrm{b}\mathscr{M}^+$, a subset $D \subseteq E$) is called *regular of degree* 2, if

$$\mathbb{E}_x[S_B^2] \ (\text{resp.}\ \mathbb{E}_\lambda[S_B^2], \sup_{y \in D} \mathbb{E}_y[S_B^2]) \text{ is finite for all } B \in \mathscr{E}^+.$$

According to the following proposition regularity of degree 2 has various equivalent characterizations. We formulate the result only for states, the result for measures and sets being similar.

Proposition 5.17. Let $x \in E$ be an arbitrary state. Each of the following six conditions is equivalent to x being regular of degree 2:

(i) $\mathbb{E}_x[S_D^2]$ is finite for some $D \in \mathscr{E}^+$, D regular of degree 2.
(ii) The measure $U_B(x, \cdot)$ is regular for all $B \in \mathscr{E}^+$.
(iii) The state x is $G_B 1$-regular for all $B \in \mathscr{E}^+$.
(iv) The measure $\bar{G}_{m_0,s,v}(x, \cdot)$ is regular.
(v) The state x is $\bar{G}_{m_0,s,v} 1$-regular.
(vi) $\bar{G}^2_{m_0,s,v}(x, E) < \infty$.

Proof. That (i) implies the regularity of degree 2 of x follows immediately from the inequality

$$S_B^2 \leq 2(S_D^2 + S_B^2 \circ \theta_{S_D}), \quad B \in \mathscr{E}.$$

If x is regular of degree 2, then by Lemma 5.4

$$U_B G_B 1(x) < \infty \quad \text{for all } B \in \mathscr{E}^+.$$

Since $U_B \geq U_A$ and $G_B \geq G_A$ whenever $B \subseteq A$, it follows that

$$U_B G_A 1(x) < \infty \quad \text{and} \quad U_A G_B 1(x) < \infty \quad \text{for all } A \supseteq B \in \mathscr{E}^+.$$

By Proposition 5.14 (ii) these imply (ii) and (iii). If, conversely, (iii) holds, then in particular

$$U_B G_B 1(x) < \infty \quad \text{for all } B \in \mathscr{E}^+.$$

But, by Lemma 5.4, this means just that x is regular of degree 2.

Now, by Proposition 5.13, (iii) is equivalent to the condition

$$\bar{G}_{m_0,s,v} G_B 1(x) < \infty \quad \text{for all } B \in \mathscr{E}^+,$$

which by Proposition 5.14 is equivalent to (iv). Similarly, (ii) and (v) are equivalent. Finally, it also follows from Proposition 5.13 that (iv), (v) and (vi) are equivalent. □

By the following proposition, if (X_n) is recurrent of degree 2 then π-almost all states $x \in E$ are regular of degree 2.

Proposition 5.18. Suppose that the Markov chain (X_n) is recurrent of degree 2. Then the set $R^2 \overset{\text{def}}{=} \{x \in E : x \text{ regular of degree } 2\}$ is equal to the full set $\{\bar{G}^2_{m_0,s,v} 1 < \infty\}$. There is an increasing sequence $D_1 \subseteq D_2 \subseteq \cdots$ of sets $D_i \in \mathscr{E}$ such that each D_i is regular of degree 2 and their union equals $\bigcup_1^\infty D_i = R^{(2)}$.

Proof. By Proposition 5.13 $\bar{G}_{m_0,s,v} 1$ is π-integrable. It follows from Proposition 5.12 that the set $\{\bar{G}^2_{m_0,s,v} 1 < \infty\}$ is full. By criterion (vi) of Proposition 5.17 this set is equal to $R^{(2)}$. Clearly, $D_i = \{\bar{G}^2_{m_0,s,v} 1 \leq i\}$, $i \geq 1$, form the desired sequence of regular sets (of degree 2). □

Examples 5.4. (*a*) Let (X_n) be a Harris recurrent, discrete Markov chain. It is recurrent of degree 2 if and only if the expectation $\mathbb{E}_x[S_x^2]$ is finite for some $x \in E$. Then $R^{(2)} = \{x \in E : \mathbb{E}_x S^2_{x_0} < \infty\}$, $x_0 \in E_\pi$ arbitrary. We have $E_\pi \subseteq R^{(2)}$.

(*d*) The reflected random walk (W_n) is recurrent of degree 2 if and only if $\mathbb{E}_0[S_0^2]$ is finite. In this case every state $x \in \mathbb{R}_+$ is regular of degree 2.

(*e*) (The forward process). Assume that F is spread-out. Then the Markov chain $(V^+_{n\delta})$ is recurrent of degree 2 if and only if $\mathbb{E}[z_1^2] = \int t^2 F(dt)$ is finite. Then every state $x \in \mathbb{R}_+$ is regular of degree 2.

5.5 Geometric recurrence

Suppose that (X_n) is a Harris recurrent Markov chain.

Definition 5.7. (i) The Markov chain (X_n) is called *geometrically recurrent* if for some small set $C \in \mathscr{S}^+$, some constant $r > 1$:

$$\sup_{x \in C} \mathbb{E}_x[r^{S_C}] < \infty. \tag{5.14}$$

(ii) A state $x \in E$ (resp. a measure $\lambda \in \mathrm{b}\mathscr{M}^+$, a subset $D \subseteq E$) is called *geometrically regular* if for all $B \in \mathscr{E}^+$ there exists a constant $r > 1$, depending on x (resp. λ, D) and on B, such that

$$\mathbb{E}_x[r^{S_B}] \text{ (resp. } \mathbb{E}_\lambda[r^{S_B}], \sup_{y \in D} \mathbb{E}_y[r^{S_B}]) \text{ is finite.}$$

Note that geometric recurrence implies positive recurrence. If there exists a geometrically regular set $D \in \mathscr{E}^+$, then (X_n) is geometrically recurrent (see Proposition 2.6).

The following proposition is the main result of this section.

Proposition 5.19. Suppose that the Markov chain (X_n) is geometrically recurrent. Then:

(i) The stationary distribution π is geometrically regular.

(ii) The small set $C \in \mathscr{S}^+$ satisfying (5.14) is geometrically regular. There exists an increasing sequence $D_1 \subseteq D_2 \subseteq \cdots$ of geometrically regular sets $D_i \in \mathscr{E}$, such that their union $\bigcup_1^\infty D_i$ is full. In particular, π-almost every state $x \in E$ is geometrically regular.

In the proof we need a couple of lemmas. Let us write

$$f_B^{(r)}(x) = \mathbb{E}_x[r^{S_B}], \quad x \in E, r \geq 1, B \in \mathscr{E}.$$

Recall the definition of the potential kernel

$$G'_B(x, A) = \mathbb{E}_x \sum_0^{S_B - 1} 1_A(X_n) = \sum_0^\infty (PI_{B^c})^n(x, A).$$

Lemma 5.5. (i) For all $r \geq 1, B \in \mathscr{E}$:

$$f_B^{(r)} = 1 + (1 - r^{-1}) G'_B f_B^{(r)}. \tag{5.15}$$

(ii) For all $B \in \mathscr{E}^+$:

$$\mathbb{E}_{\pi I_B}[r^{S_B}] = \pi(B) + (1 - r^{-1}) \mathbb{E}_\pi[r^{S_B}].$$

(iii) If there exists a geometrically regular set $D \in \mathscr{E}^+$ then π is geometrically regular.

Proof. (i) We have

$$(1 - r^{-1})^{-1}(f_B^{(r)}(x) - 1) = \mathbb{E}_x \sum_1^{S_B} r^n = \mathbb{E}_x \sum_0^{S_B - 1} r^{S_B - n}$$

$$= \mathbb{E}_x \sum_0^{S_B - 1} \mathbb{E}_{X_n} r^{S_B},$$

by conditioning w.r.t. $\mathscr{F}_n^X$ and using the fact that $S_B - n = S_B \circ \theta_n$ on $\{S_B > n\}$ for $n \geq 0$,

$$= G'_B f_B^{(r)}(x).$$

(ii) Integrate both sides of the identity (5.15) over B w.r.t. π and use Proposition 5.9.

(iii) By (ii), and since $S_B \le S_D + S_B \circ \theta_{S_D}$, we have for all $B \in \mathscr{E}^+$:

$$\mathbb{E}_\pi[r^{S_B}] \le \mathbb{E}_\pi[r^{S_D} r^{S_B \circ \theta_{S_D}}]$$
$$\le (1 - r^{-1})^{-1} \sup_{x \in D} \mathbb{E}_x[r^{S_D}] \sup_{y \in D} \mathbb{E}_y[r^{S_B}].$$

Since D is geometrically regular, the right hand side is finite for sufficiently small $r > 1$. □

The second lemma is a general result concerning the growth of the expectations of random sums (cf. Lemmas 5.1 and 5.2):

Lemma 5.6. Suppose that $(\xi_n; n \ge 0)$ is a non-negative stochastic process, adapted to a history $(\mathscr{F}_n; n \ge 0)$. Let τ be a stopping time relative to $(\mathscr{F}_n)$. Suppose that for some constant $\gamma > 0$,

$$\mathbb{P}\{\tau = n \mid \mathscr{F}_{n-1}\} \ge \gamma \quad \text{on } \{\tau \ge n\} \quad \text{for all } n \ge 1, \tag{5.16}$$

and that for some constants $r_0 > 1$ and $M_0 < \infty$,

$$\mathbb{E}[r_0^{\xi_n} \mid \mathscr{F}_{n-1}] \le M_0 \quad \text{for all } n \ge 1. \tag{5.17}$$

Then there exist constants $r_1 > 1$ and $M_1 < \infty$ such that

$$\mathbb{E}[r^{\sum_0^\tau \xi_n}] \le M_1 \mathbb{E}[r^{\xi_0}] \quad \text{for all } 1 \le r \le r_1.$$

Proof. It follows from (5.17) that

$$\mathbb{P}\{\xi_n \ge m \mid \mathscr{F}_{n-1}\} \le M_0 r_0^{-m} \quad \text{for all } m \ge 0, n \ge 1.$$

Put $f(r) = 1 + (r-1)\sum_1^\infty r^{m-1} q_m$, where $q_m = \min\{M_0 r_0^{-m}, 1\}$. Then $f(r)$ is a probability generating function satisfying

$$\mathbb{E}[r^{\xi_n} \mid \mathscr{F}_{n-1}] \le f(r) \quad \text{for all } n \ge 1, 1 \le r \le r_0. \tag{5.18}$$

Consequently we have on $\{\tau \ge 1\}$ for all $1 \le r \le r_0$ (using the notation $\mathbb{E}^{(n)}$ for the conditional expectation $\mathbb{E}[\cdot \mid \mathscr{F}_n]$):

$$\mathbb{E}^{(0)}[r^{\sum_1^\tau \xi_n}] \le \sum_{N=1}^\infty \mathbb{E}^{(0)}[r^{\sum_1^N \xi_n}; \tau \ge N]$$
$$\le f(r) \sum_{N=1}^\infty \mathbb{E}^{(0)}[r^{\sum_1^{N-1} \xi_n}; \tau \ge N] \tag{5.19}$$

by conditioning w.r.t. $\mathscr{F}_{N-1}$. Further, by using the Schwarz inequality, (5.16) and (5.18), we obtain for all $N \ge 2$ and $1 \le r \le (r_0)^{1/2}$:

$$\mathbb{E}^{(0)}[r^{\sum_1^{N-1} \xi_n}; \tau \ge N]$$
$$= \mathbb{E}^{(0)}[\mathbb{E}^{(N-2)}[r^{\xi_{N-1}}; \tau \ne N-1] r^{\sum_1^{N-2} \xi_n}; \tau \ge N-1]$$
$$\le (1-\gamma)^{1/2} (f(r^2))^{1/2} \mathbb{E}^{(0)}[r^{\sum_1^{N-2} \xi_n}; \tau \ge N-1].$$

Hence by induction

$$\mathbb{E}^{(0)}[r^{\Sigma_1^{N-1}\xi_n}; \tau \geq N] \leq (1-\gamma)^{(N-1)/2}(f(r^2))^{(N-1)/2} \quad \text{for all } N \geq 2.$$

Substituting this into (5.19) gives

$$\mathbb{E}^{(0)}[r^{\Sigma_1^{\tau}\xi_n}] \leq f(r) \sum_{N=1}^{\infty} (1-\gamma)^{(N-1)/2}(f(r^2))^{(N-1)/2} \quad \text{on } \{\tau \geq 1\}.$$

Since $f(r) \downarrow 1$ as $r \downarrow 1$, the right hand side of this inequality is finite, say M_1, for some $r_1 > 1$.

The proof is completed after observing that

$$\begin{aligned}\mathbb{E}[r^{\Sigma_0^{\tau}\xi_n}] &= \mathbb{E}[r^{\xi_0}; \tau = 0] + \mathbb{E}[r^{\xi_0}; \tau \geq 1; \mathbb{E}^{(0)}[r^{\Sigma_1^{\tau}\xi_n}]] \\ &\leq M_1 \mathbb{E}[r^{\xi_0}] \quad \text{for all } 1 \leq r \leq r_1. \quad \square\end{aligned}$$

Proof of Proposition 5.19. First we shall show that the small set C is geometrically regular.

Let $B \in \mathscr{E}^+$ be arbitrary. Since C is small there exists an integer $n_0 \geq 1$ such that

$$\gamma = \inf_{y \in C} P^{n_0}(y, B) > 0,$$

(cf. Proposition 2.7(i)). Set $\eta(0) = T_C$, $\eta(i) = \inf\{n \geq \eta(i-1) + n_0 : X_n \in C\}$ for $i \geq 1$, $(\mathscr{F}_i) = (\mathscr{F}^X_{\eta(i)})$, $\xi_0 = \eta(0)$, $\xi_i = \eta(i) - \eta(i-1)$ for $i \geq 1$, and define a stopping time τ relative to $(\mathscr{F}_i)$ by

$$\tau = \inf\{i \geq 1 : X_{\eta(i-1)+n_0} \in B\}.$$

By using Lemma 5.6 we can conclude that

$$\sup_{x \in C} \mathbb{E}_x r^{S_B} < \infty \quad \text{for some } r > 1;$$

i.e., C is geometrically regular.

Now, by Lemma 5.5(iii), π is geometrically regular.

In particular, $\mathbb{E}_\pi r_0^{S_C}$ is finite for some $r_0 > 1$. Set

$$D_i = \{x \in E : \mathbb{E}_x r_0^{S_C} \leq i\}, \quad i \geq 1.$$

Then $\bigcup_1^\infty D_i$ is full. By using the inequality $S_B \leq S_C + S_B \circ \theta_{S_C}$ we get for all $i \geq 1, x \in D_i, B \in \mathscr{E}^+, 1 \leq r \leq r_0$:

$$\mathbb{E}_x r^{S_B} \leq i \sup_{y \in C} \mathbb{E}_y r^{S_B}.$$

The right hand side is finite for sufficiently small $r > 1$, because C is geometrically regular. Thus each D_i is geometrically regular. $\square$

Suppose that P satisfies the minorization condition $M(m_0, 1, s, \nu)$ for some $m_0 \geq 1, s \in \mathscr{S}^+$, $\nu \in \mathscr{M}^+$. In the following proposition geometric recurrence is characterized in terms of the potential kernel $G^{(r)}_{m_0,s,\nu} = \sum_0^\infty r^{nm_0}(P^{m_0} - s \otimes \nu)^n$. In order to avoid unessential technicalities we consider the aperiodic case only.

Proposition 5.20. Suppose that (X_n) is aperiodic. Then (X_n) is geometrically recurrent if and only if $\nu G^{(r)}_{m_0,s,\nu} s$ is finite for some $r > 1$.

Remark 5.1. If $m_0 = 1$, i.e., (s, ν) is an atom, we have $r\nu G^{(r)}_{1,s,\nu} s = r\nu G^{(r)}_{s,\nu} s = \mathbb{E}_\alpha r^{S_\alpha}$ (see (4.20)).

In the proof of Proposition 5.20 we need the following:

Lemma 5.7. If (X_n) is aperiodic and geometrically recurrent then so is the m_0-step chain (X_{nm_0}).

Proof. Let $C \in \mathscr{E}^+$ be a small set satisfying (5.14). Set $B = C$, $f \equiv 1$, and define the history $(\mathscr{F}_i)$, the stochastic process (ξ_i) and the stopping time τ as in the proof of Lemma 5.3. By using Lemma 5.6 we obtain the desired result

$$\sup_{x \in C} \mathbb{E}_x [r^{m_0 S_C}\rangle < \infty. \qquad \square$$

Proof of Proposition 5.20. Suppose that (X_n) is geometrically recurrent. By Lemma 5.7 there is no loss of generality in assuming that $m_0 = 1$. Now it is easy to see (using Lemma 5.6 in an obvious manner) that in this case the split chain (X_n, Y_n), too, is geometrically recurrent. Now the finiteness of $\mathbb{E}_\alpha r^{S_\alpha}$ for some $r > 1$ follows from Proposition 5.19.

The converse result is as easy. Since, trivially, the geometric recurrence of (X_{nm_0}) implies that of (X_n), we can again consider only the case $m_0 = 1$. By Remark 5.1 the condition $\nu G^{(r)}_{s,\nu} s < \infty$ implies that the split chain (X_n, Y_n) is geometrically recurrent. By applying Proposition 5.19 we see that there exists a geometrically regular set $D \in \mathscr{E}^+$ (for (X_n)). But this implies that (X_n) is geometrically recurrent. $\square$

The following proposition gives a characterization for geometric recurrence in terms of a drift condition (cf. Proposition 5.10):

Proposition 5.21. Suppose that for some small set $C \in \mathscr{S}^+$, function $g \in \mathscr{E}_+$ and constant $r > 1$:

$$\sup_{x \in C^c} \mathbb{E}[rg(X_{n+1}) - g(X_n) | X_n = x] = \sup_{C^c}(rPg - g) < 0 \tag{5.20}$$

and

$$\sup_{x \in C} \mathbb{E}[g(X_{n+1}); X_{n+1} \in C^c | X_n = x] = \sup_C P I_{C^c} g < \infty.$$

Then the Markov chain (X_n) is geometrically recurrent.

Proof. From the inequality (5.20) we can derive – precisely in the same manner as we did in the proof of Proposition 5.10 – the following:

$$g(x) \geq \gamma 1_{C^c}(x)\mathbb{E}_x \sum_{0}^{S_C - 1} r^n, \quad \gamma > 0 \text{ a constant.}$$

Hence by our hypothesis

$$\infty > \sup_C PI_{C^c}g \geq \gamma \sup_C \mathbb{E}_x\left[\sum_{0}^{S_C - 2} r^n ; S_C \geq 2\right],$$

from which the assertion easily follows. □

Examples 5.5. (*a*) Let (X_n) be a Harris recurrent, discrete Markov chain. It is geometrically recurrent if and only if $\mathbb{E}_x[r^{S_x}]$ is finite for some state $x \in E$ and some constant $r = r(x) > 1$. In this case for all $x \in E_\pi$, $E_\pi[r_0^{S_x}]$ is finite for some constant $r_0 = r_0(x) > 1$.

(*d*) The reflected random walk (W_n) is geometrically recurrent if $\mathbb{E}z_1 < 0$ and $\mathbb{E}e^{\gamma z_1} < \infty$ for some $\gamma > 0$. (Hint: Set $g(x) = e^{\varepsilon x}$, $C = [0, c]$, for sufficiently small $\varepsilon > 0$ and big $0 < c < \infty$, in Proposition 5.21.)

(*e*) (The forward process). Suppose that F is spread out. The Markov chain $(V_{n\delta}^+)$ is geometrically recurrent if and only if $\mathbb{E}e^{\gamma z_1} = \int e^{\gamma t} F(dt)$ is finite for some constant $\gamma > 0$.

(*f*) The autoregressive process $R_n = \rho R_{n-1} + z_n, n \geq 1$, with $|\rho| < 1$, $F = \mathscr{L}(z_1)$ non-singular, and $\int |t| F(dt) = \mathbb{E}|z_1|$ finite, is geometrically recurrent. (Hint: Set $g(x) = |x|$, $C = [-c, c]$, c big enough, in Proposition 5.21.)

5.6 Uniform recurrence

The strongest from of recurrence we consider is uniform recurrence.

We start with a brief discussion on special sets. Let (X_n) be an irreducible Markov chain. In Definition 5.4 we introduced the concepts of a special function and set for Harris recurrent (X_n). Let us extend this definition to arbitrary irreducible (X_n). So, for example, a set $D \in \mathscr{E}^+$ is called *special*, if

$$\sup_E U_B 1_D = \sup_{x \in E} \mathbb{E}_x \sum_{n=1}^{S_B} 1_D(X_n) < \infty \quad \text{for all } B \in \mathscr{E}^+. \tag{5.21}$$

It turns out that special sets are 'test sets' for recurrence and positive recurrence (cf. also Theorem 3.7(vii) and Corollary 5.3(ii)):

Proposition 5.22. Suppose that there exists a special set $D \in \mathscr{E}^+$ satisfying

$$U_D(x, D) = \mathbb{P}_x\{S_D < \infty\} = 1 \quad \text{for all } x \in D. \tag{5.22}$$

Then the Markov chain (X_n) is recurrent. It is Harris recurrent on the absorbing set $D^\infty = \{h_D = 1\} = \{x \in E : \mathbb{P}_x\{T_D < \infty\} = 1\}$. If, in addition,

$$\sup_{x \in D} \mathbb{E}_x S_D < \infty \tag{5.23}$$

then (X_n) is positive recurrent.

Proof. From the hypothesis (5.22) it follows that $X_n \in D$ i.o. $\mathbb{P}_x$-a.s. for all $x \in D$. If it were true that $\mathbb{P}_x\{S_B = \infty\} > 0$ for some $x \in D^\infty$, $B \in \mathscr{E}^+$, then this would lead to a contradiction with (5.21). Consequently, (X_n) is Harris recurrent on D^∞.

To see that (5.23) implies positive recurrence, take a small set $C \in \mathscr{S}^+$ with $C \subseteq D$. From the resolvent equation (3.18) it follows that

$$\sup_D U_C 1 = \sup_D (U_D 1 + U_C I_{D \setminus C} U_D 1) \leq (1 + \sup_E U_C 1_D) \sup_D U_D 1 < \infty.$$

By Corollary 5.3(ii), (X_n) is positive recurrent. □

We state the following:

Definition 5.8. An irreducible Markov chain (X_n) is called *uniformly recurrent* if

$$\sup_{x \in E} \mathbb{E}_x S_B < \infty \quad \text{for all } B \in \mathscr{E}^+.$$

Note that uniform recurrence implies Harris recurrence; then, in particular, the transition probability P is stochastic.

Note also that uniform recurrence coincides with the requirements that P be stochastic and the state space E be special.

Other characterizations are given in the following:

Proposition 5.23. Suppose that P is stochastic and irreducible. Each of the following three conditions is equivalent to uniform recurrence:

(i) (X_n) is uniformly geometrically recurrent in the sense that for all $B \in \mathscr{E}^+$ there exists a constant $r = r(B) > 1$ such that

$$\sup_{x \in E} \mathbb{E}_x[r^{S_B}] < \infty.$$

(ii) $\sup_{x \in E} \mathbb{E}_x S_D < \infty$ for some special set $D \in \mathscr{E}^+$.

(iii) $\inf_{x \in E} \sum_1^N P^n(x, D) > 0$ for some integer $N \geq 1$, some special $D \in \mathscr{E}^+$.

Proof. Clearly (i) implies uniform recurrence. Conversely, if (X_n) is uniformly recurrent then for any $B \in \mathscr{E}^+$,

$$\sup_{x \in E} \mathbb{P}_x\{S_B > N\} < 1 \tag{5.24}$$

for N large enough. It is easy to see that this implies (i).

By Proposition 5.22, if (ii) holds then (X_n) is positive Harris recurrent. By Proposition 5.14(i), in order to prove that (X_n) is uniformly recurrent it suffices to show that $\mathbb{E}_x S_B = U_B 1(x)$ is bounded for some regular set $B \in \mathscr{E}^+ \cap D$. By (3.18)

$$\sup_E U_B 1 = \sup_E (U_D 1 + U_B I_{D \setminus B} U_D 1)$$
$$\leq (1 + \sup_E U_B 1_D) \sup_E U_D 1 < \infty .$$

That (iii) follows from uniform recurrence is a direct consequence of (5.24). Conversely, if (iii) holds then there is a constant $\delta > 0$ and a finite partition $E = \sum_1^N E_n$ of E such that for all $1 \leq n \leq N$ and $x \in E_n$,

$$P^n(x, D) \geq \delta.$$

Hence

$$\sup_{x \in E} \mathbb{P}_x\{S_D > N\} \leq 1 - \delta,$$

implying (ii). □

Using (ii) and Proposition 5.13(ii) we see that uniform recurrence is equivalent to the condition $\sup_E \mathbb{E}_x S_B < \infty$ for all $B \in \mathscr{E}^+ \cap B_0$, for some $B_0 \in \mathscr{E}^+$.

The 'drift criterion' now takes the following form:

Proposition 5.24. Suppose that P is stochastic and irreducible, and that for some bounded, non-negative function $g \in b\mathscr{E}_+$, for some special (in particular, some small) set $D \in \mathscr{E}^+$:

$$\sup_{D^c} (Pg - g) < 0.$$

Then the Markov chain (X_n) is uniformly recurrent.

Proof. Use Propositions 5.10 and 5.23(ii). □

Examples 5.6. (*a*) Let (X_n) be a discrete Markov chain. It is uniformly recurrent if and only if $\sup_{x \in E} \mathbb{E}_x S_y$ is finite for some $y \in E$. Then this quantity is finite for all $y \in E_\pi$. Every irreducible Markov chain on a *finite* state space E and with a stochastic transition matrix P is uniformly recurrent.

(*e*) (The forward process). The Markov chain $(V_{n\delta}^+)$ is uniformly recurrent if and only if $\bar{M} = \text{ess sup } z_1 = \sup\{t : F(t) < 1\}$ is finite.

(*f*) (The autoregressive process). Suppose that $|\rho| < 1$. Also suppose that F is non-singular and there is a finite interval $[a, b]$, $-\infty < a < b < \infty$, such that F is supported by $[a, b]$. Then the autoregressive process (R_n) is uniformly recurrent on the absorbing set $\Gamma = [(1-\rho)^{-1} a, (1-\rho)^{-1} b]$.

(*i*) The storage chain (S_n) introduced in Example 4.1(*i*) is uniformly recurrent.

We shall introduce still one more example:

(*k*) Let (X_n) be an irreducible and aperiodic Markov chain with convergence parameter $R=1$. Suppose that for some integers n_0, $I \geq 1$, constant $M<\infty$, small function $s_1,\dots,s_I \in \mathscr{S}^+$ and small measures $\nu_1,\dots,\nu_I \in \mathscr{M}^+$ the following *majorization condition* holds:

$$P^{n_0} \leq M \sum_{i,j=1}^{I} s_i \otimes \nu_j.$$

Then (X_n) is recurrent. It is uniformly recurrent on every Harris set $H \in \mathscr{E}$. (For hints for the proof see the forthcoming Example 5.7(*k*).)

5.7 Degrees of R-recurrence

In this section we shall extend some of the concepts of the previous sections to general kernels. Our analysis will not, however, be as thorough as in the case of a transition probability P. Instead, we shall only briefly discuss the extensions of the concepts of geometric and uniform recurrence.

We assume that K is an R-recurrent kernel satisfying the minorization condition $M(m_0,1,s,\nu)$. We let π denote an R-invariant measure for K. We assume that $h=h_\nu$ is the unique minimal R-invariant function satisfying $\nu(h)=1$. Then by Proposition 5.4, the transformed kernel

$$\tilde{K} = R I_{h^{-1}} K I_h$$

is the transition probability of a Harris recurrent Markov chain $(\tilde{X}_n)$ on $\tilde{E}=\{h<\infty\}$. $(\tilde{X}_n)$ is positive recurrent if and only if K is R-positive recurrent.

For any $B \in \mathscr{E}$, $r \geq 0$, let $U_B^{(r)}$ denote the kernel defined by

$$U_B^{(r)} = \sum_1^\infty r^n K(I_{B^c}K)^{n-1},$$

and let $\tilde{U}_B^{(r)}$ denote the kernel

$$\tilde{U}_B^{(r)} = \sum_1^\infty r^n \tilde{K}(I_{B^c}\tilde{K})^{n-1}.$$

Clearly,

$$\tilde{U}_B^{(r)} = I_{h^{-1}} U_B^{(rR)} I_h \quad \text{for all } r \geq 0.$$

Hence

$$\mathbb{E}_x[r^{\tilde{S}_B}] = \tilde{U}_B^{(r)}(x,B) = (h(x))^{-1} U_B^{(rR)} I_h(x,B). \tag{5.25}$$

We state the following:

Definition 5.9. An R-recurrent kernel K is called *geometrically R-recurrent,* if for some small set $C \in \mathscr{S}^+$ such that $\inf_C h > 0$ and $\sup_C h < \infty$, and for some constant $r > R$:

$$\sup_C U_C^{(r)} 1_C < \infty.$$

If h is bounded on a small set $C \in \mathscr{S}^+$, then C is small for the transformed kernel $\tilde{K}$, too. From (5.25), Definition 5.7 and Proposition 5.19(ii) we easily get the following:

Proposition 5.25. The kernel K is geometrically R-recurrent if and only if the Markov chain $(\tilde{X}_n)$ is geometrically recurrent. □

As a direct consequence of Proposition 5.20 and the above result we have:

Corollary 5.6. Suppose that K is aperiodic. Then K is geometrically R-recurrent if and only if $\nu G_{m_0,s,\nu}^{(r)} s$ is finite for some $r > R$. □

Let $G_B^{(r)}$ denote the potential kernel of $r(I_{B^c}K)$, i.e.,

$$G_B^{(r)} = \sum_0^\infty r^n (I_{B^c}K)^n.$$

We shall call a π-integrable function $g \in \mathscr{L}^1(\pi)$ *special*, if for every $B \in \mathscr{E}^+$ there is a constant $M = M(B) < \infty$ such that

$$G_B^{(R)}|g| \le Mh. \tag{5.26}$$

Since (using obvious notation)

$$\tilde{G}_B = I_{h^{-1}} G_B^{(R)} I_h,$$

we obtain the following result (see Corollary 5.5(ii)):

Proposition 5.26. (i) A function $g \in \mathscr{L}^1(\pi)$ is special for K if and only if the function $\tilde{g} = h^{-1}g$ is special for $(\tilde{X}_n)$.

(ii) A function $g \in \mathscr{L}^1(\pi)$ is special for K if (and only if) there is a set $B_0 \in \mathscr{E}$ such that (5.26) holds for all $B \in \mathscr{E}^+ \cap B_0$. □

We state the following:

Definition 5.10. An R-recurrent kernel K is *uniformly R-recurrent* if the R-invariant function h is special; i.e. if for every $B \in \mathscr{E}^+$ there is a constant $M = M(B) < \infty$ such that

$$G_B^{(R)} h \le Mh. \tag{5.27}$$

As an immediate consequence of Propositions 5.23 and 5.26 we have (see also the remark made after Proposition 5.23):

Corollary 5.7. Each of the following three conditions is equivalent to uniform R-recurrence:

(i) The Markov chain $(\tilde{X}_n)$ is uniformly recurrent.

(ii) For some $B_0 \in \mathscr{E}^+$: the inequality (5.27) holds for all $B \in \mathscr{E}^+ \cap B_0$.

(iii) For some $B_0 \in \mathscr{E}^+$: for all $B \in \mathscr{E}^+ \cap B_0$ there exist an integer $N = N(B)$ and a constant $\gamma = \gamma(B) > 0$ such that

$$\sum_1^N K^n 1_B \geq \gamma h \quad \text{on } \{h < \infty\}. \qquad \square$$

Remark 5.2. In fact, for Corollary 5.7 we need not assume that K is R-recurrent, but only that K is irreducible and has an R-invariant function h on some closed set F. This holds true since then the transformed kernel $\tilde{K} = RI_{h^{-1}}KI_h$ is stochastic on the closed set $F \cap \{h < \infty\}$ (cf. Proposition 5.23).

Examples 5.7. (*a*) Let K be a Card_F-irreducible matrix, Card_F a maximal irreducibility measure. Fix a state $z \in F$ and set

$$b_n = K({}_zK)^{n-1}(z, z), \quad n \geq 1,$$

(see Example 4.1(*a*)). The kernel K is geometrically R-recurrent if and only if

$$\hat{b}(r) = \sum_{n=1}^{\infty} r^n b_n < \infty \quad \text{for some } r > R.$$

Suppose that K has an R-invariant column vector $h = (h(x); x \in E)$. K is uniformly R-recurrent if for some $z \in F, N \geq 1, \gamma > 0$:

$$\sum_1^N k^n(x, z) \geq \gamma h(x) \quad \text{whenever } h(x) < \infty.$$

In particular, every irreducible (non-negative) matrix on a *finite* state space is uniformly R-recurrent.

(*k*) Let K be an irreducible aperiodic kernel. Suppose that for some integers $n_0, I \geq 1$, constant $M < \infty$, small functions $s_1, \ldots, s_I \in \mathscr{S}^+$ and small measures $v_1, \ldots, v_I \in \mathscr{M}^+$ the following *majorization condition* holds:

$$K^{n_0} \leq M \sum_{i,j=1}^{I} s_i \otimes v_j.$$

Then K is uniformly R-recurrent.

Hints for the proof: First note that, by Proposition 2.10(iii), we may suppose that

$$K^{n_0} \leq M s \otimes v$$

for some single small function s and single small measure v satisfying the minorization condition $M(m_0, 1, s, v)$. By considering suitable iterates of K we can conclude that there is no loss of generality in assuming that $m_0 = n_0$. Further, since the 'R-properties' of K are inherited by K^{m_0} as the corresponding 'R^{m_0}-properties' we may assume that $m_0 = n_0 = 1$; i.e., finally we are led to the simple hypothesis

$$s \otimes v \leq K \leq Ms \otimes v.$$

Let $0 \leq R < \infty$ be the convergence parameter of K. By Proposition 4.7

$$R = \sup\{r \geq 0 : \sum_0^\infty r^{n+1} v(K - s \otimes v)^n s < 1\}$$

$$= \sup\{r \geq 0 : \sum_0^\infty r^n v K^n s < \infty\}.$$

Since by our hypothesis

$$K - s \otimes v \leq \rho K, \quad \text{where } \rho = 1 - M^{-1} < 1,$$

it follows that the series $rvG_{s,v}^{(r)}s = \sum_0^\infty r^{n+1} v(K - s \otimes v)^n s$ converges for $r = \rho^{-1/2} R > R$; by Corollary 5.6 this implies that K is geometrically R-recurrent. Let $h = h_v$ be the unique minimal R-invariant function. Note that $h > 0$ everywhere. We have to show that h is special. By our hypothesis

$$h = RKh \leq Mv(h)s.$$

Hence h is small, and by Propositions 5.13(iii) and 5.26, special.

Note that the Markov chain of Example 5.6(k) is uniformly 1-recurrent.

6

Total variation limit theorems

In this chapter we shall study for an R-recurrent kernel K the convergence of the iterates K^n as $n \to \infty$.

The main result here is Orey's convergence theorem. Formulated for an aperiodic Harris recurrent Markov chain with transition probability P, this theorem states that the n-step transition probabilities $P^n(x, \cdot)$ converge in total variation norm: for any two initial distributions λ and μ,

$$\lim_{n \to \infty} |\lambda P^n - \mu P^n|(E) = 0. \tag{6.1}$$

For a general aperiodic R-recurrent kernel K, having the (essentially unique) minimal R-invariant function h, Orey's theorem takes the following form: for any two measures $\lambda, \nu \in \mathcal{M}^+$ such that $\lambda(h) = \mu(h) < \infty$,

$$\lim_{n \to \infty} R^n |\lambda K^n - \mu K^n|(h) = 0. \tag{6.2}$$

We shall also consider various sharpenings to Orey's theorem by considering the rate with which the norms in (6.1) and (6.2) tend to zero. We shall see that the rate of convergence in Orey's theorem is closely related to the degree of recurrence of the Markov chain (X_n) (or of the kernel K).

The basic technique we shall use in the proofs is the regeneration method introduced in Chapter 4. Henceforth we assume that $m_0 \geq 1$, $s \in \mathcal{S}^+$ and $\nu \in \mathcal{M}^+$ are fixed such that the minorization condition $M(m_0, 1, s, \nu)$ holds. When $m_0 = 1$, (X_n, Y_n) denotes the split chain induced by the atom (s, ν).

Since the regeneration method is based on the exploitation of the embedded renewal process, we start by studying renewal theory, i.e. the asymptotic behaviour of renewal sequences.

6.1 Renewal theory

We adopt the notation and terminology used in Section 4.1. Suppose that $(T(i);\ i \geq 0)$ is a recurrent renewal process with increment distribution b. If $a = (a_n;\ n \geq 0)$ denotes a delay distribution then the corresponding (delayed) renewal sequence $v = (v_n;\ n \geq 0)$ is given by $v = a * u$. Recall the definition of the sequence $B_n = \sum_{n+1}^{\infty} b_m, n \geq 0$. We have $u * B \equiv 1$ (see (4.6)).

The basic renewal theorem is:

Theorem 6.1. Suppose that $u=(u_n;n\geq 0)$ is an aperiodic, recurrent renewal sequence. Let $a=(a_n;n\geq 0)$ be an arbitrary delay distribution. Then

$$\lim_{n\to\infty} |a*u-u|*B_n=0.$$

The proof of this theorem is based on the use of the so-called *coupling technique*. It involves the study of two renewal processes defined on the same probability space, and having the same increment distribution but different delays. Their joint distribution is constructed in such a manner that they eventually have a common renewal epoch (a.s.). This leads to an inequality, called the coupling inequality, from which the convergence result easily follows.

In proving that the coupling time is finite we need the following lemma. Its proof can be found e.g. in Feller (1971), Sect. VI.10, Theorem 4.

Lemma 6.1. Let $Z=(Z(i);i\geq 0)$ be an aperiodic, integer valued random walk; i.e.

$$Z(i)=Z(0)+\sum_1^i z(j),\quad i\geq 1,$$

where $Z(0)$ is an integer valued random variable, and the increments $z(j)$ are i.i.d., independent of $Z(0)$, integer valued random variables having a common non-lattice distribution. Suppose that the increments have zero expectation, $\mathbb{E}z(1)=0$. Then the random walk Z is an aperiodic and recurrent Markov chain on $E=\cdots,-1,0,1,2,\ldots$. □

Proof of Theorem 6.1. Assume first that, instead of mere aperiodicity, the following stronger condition holds true:

$$\text{g.c.d.}\{n-m\colon m<n, b_m>0, b_n>0\}=1.$$

Let $N\geq 1$ be an integer so big that

$$\text{g.c.d.}\{n-m\colon m<n\leq N, b_m>0, b_n>0\}=1. \tag{6.3}$$

Let $T=(T(i);i\geq 0)$, $T(0)=0$, $T(i)=\sum_1^i t(j)$ for $i\geq 1$, be an undelayed renewal process with increment distribution b. We shall construct a sequence $T'=(T'(i);i\geq 0)$ of random times, $T'(i)=T'(0)+\sum_1^i t'(j)$ for $i\geq 1$, in the following manner:

Set $T'(0)=1$. For any $j\geq 1$, if the increment $t(j)$ is bigger than N, then we set $t'(j)=t(j)$, whereas if $t(j)$ is smaller than or equal to N, then we distribute $t'(j)$ independently of, and with the same (conditional) distribution as, $t(j)$. In exact terms, for every $j\geq 1$,

$$\mathbb{P}\{t'(j)=t(j)|\mathscr{F}_{j-1}^{T'}\vee\mathscr{F}_j^T\}=1\quad \text{on } \{t(j)>N\}$$

and

$$\mathscr{L}(t'(j)|\mathscr{F}^{T'}_{j-1} \vee \mathscr{F}^{T}_{j}) = \mathscr{L}(t(j)|t(j) \leq N)$$
$$= (1-B_N)^{-1} b_{(\cdot)} \quad \text{on } \{t(j) \leq N\}.$$

It is clear that the marginal probability law of T' is the law of a delayed renewal process with delay $T'(0) = 1$ and with increment distribution b.

By (6.3) and by the above construction the random walk Z defined by

$$Z(i) = T'(i) - T(i), \quad i \geq 0,$$

has symmetric, bounded (by N), non-lattice increments $z(j) = t'(j) - t(j)$. Hence it satisfies the hypotheses of Lemma 6.1. It follows that the random time

$$\eta = \inf\{i \geq 1 : Z(i) = 0\} = \inf\{i \geq 1 : T'(i) = T(i)\}$$

is finite almost surely.

Let $V = (V_n; n \geq 0)$ and $V' = (V'_n; n \geq 1)$ be the backward chains associated with the renewal processes T and T', respectively. (See (4.10). Note that V'_0 is not defined.) The initial states of the backward chains V and V' are $V_0 = 0$ and $V'_1 = 0$, respectively. Note that, since the increments of T and T' have the same distribution b, the Markov chains $V = (V_n; n \geq 0)$ and $(V'_{1+n}; n \geq 0)$ obey the same probability law.

Note that the random time $\tau = T(\eta) = T'(\eta)$ is a randomized stopping time for both the Markov chains V and V'.

Let $A \subseteq \mathbb{N}$ be arbitrary. By using the Markov property at $\tau = m$ and the fact that $V_\tau = V'_\tau = 0$, we obtain for any $n \geq 1$:

$$\mathbb{P}\{V_n \in A\} = \sum_{m=1}^{n} \mathbb{P}\{V_{n-m} \in A\} \mathbb{P}\{\tau = m\} + \mathbb{P}\{V_n \in A, \tau > n\},$$

$$\mathbb{P}\{V'_n \in A\} = \sum_{m=1}^{n} \mathbb{P}\{V_{n-m} \in A\} \mathbb{P}\{\tau = m\} + \mathbb{P}\{V'_n \in A, \tau > n\}.$$

From these we get the *coupling inequality*

$$\sup_{A \subseteq \mathbb{N}} |\mathbb{P}\{V_n \in A\} - \mathbb{P}\{V'_n \in A\}| \leq \mathbb{P}\{\tau > n\}.$$

Since τ is finite a.s., the right hand side tends to zero as $n \to \infty$. The left hand side is equal to

$$\frac{1}{2}\sum_{m=0}^{\infty} |\mathbb{P}\{V_n = m\} - \mathbb{P}\{V'_n = m\}| = \frac{1}{2}\sum_{m=0}^{n} |u_{n-m}B_m - u_{n-1-m}B_m|$$
$$= \tfrac{1}{2}|u - \delta^{(1)} * u| * B_n$$

where $\delta^{(1)}$ is the probability distribution on $\mathbb{N}$ assigning unit mass to the integer 1. Consequently,

$$\lim_{n \to \infty} |u - \delta^{(1)} * u| * B_n = 0. \tag{6.4}$$

Let us now drop the additional assumption (6.3) and assume only the aperiodicity, i.e.

$$\text{g.c.d.}\{n \geq 1 : b_n > 0\} = 1.$$

Note that proceeding as above would lead to a possibly periodic random walk Z. In order to deal with this complication, we use the following trick:

Let $0 < p < 1$ be a constant. We modify the increment distribution b by setting

$$\bar{b}_0 = p, \bar{b}_n = (1-p)b_n \quad \text{for } n \geq 1 \quad (\text{or, briefly, } \bar{b} = p\delta + (1-p)b).$$

Let $(\bar{T}(i); i \geq 0)$ be the associated renewal process. (Note that we allow now, exceptionally, an increment to be zero with positive probability; cf. (4.1).) It is easy to see that the above proof goes through for the modified process. Hence we get (6.4) for the modified renewal sequence $\bar{u}$. But since

$$\bar{u} = (1-p)^{-1}u \quad \text{and } \bar{B} = (1-p)B$$

we obtain (6.4) for the original renewal sequence u, too.

If we have an arbitrary distribution a in place of $\delta^{(1)}$, we proceed as follows. For any $N \geq 0$, let $a^{(N)}$ denote the truncated sequence

$$\begin{aligned} a_n^{(N)} &= a_n \quad \text{for } 0 \leq n \leq N, \\ &= 0 \quad \text{for } n > N. \end{aligned}$$

Write

$$A_N = \sum_{N+1}^{\infty} a_n.$$

We have for all $n \geq 0$:

$$\begin{aligned} |u - a*u| * B_n \leq & \left| \left(\sum_0^N a_m \right) u - a^{(N)} * u \right| * B_n \\ & + A_N u * B_n + (a - a^{(N)}) * u * B_n. \end{aligned}$$

By (6.4), for any fixed N, the first term on the right hand side tends to zero as $n \to \infty$. By (4.6), the second and third terms are both dominated by A_N, and they can therefore be made arbitrarily small by choosing N big enough. □

Let $a = (a_n; n \geq 0)$ be an arbitrary non-negative sequence. Set

$$M_a^{(0)} = \sum_0^{\infty} a_n, \quad M_a = \sum_0^{\infty} n a_n.$$

Note that $M_b = \mathbb{E}t$ is finite or infinite depending on whether the renewal process $(T(i); i \geq 0)$ is positive or null recurrent.

From Theorem 6.1 we are able to deduce the following:

Theorem 6.2. Suppose that u is an aperiodic, recurrent renewal sequence. Then

$$\lim_{n\to\infty} a*u_n = M_b^{-1}M_a^{(0)}$$

unless $M_a^{(0)} = M_b = \infty$. In particular, the limit

$$u_\infty = \lim_{n\to\infty} u_n = M_b^{-1}$$

exists; $u_\infty > 0$ or $= 0$ depending on whether the renewal sequence u is positive or null recurrent.

Proof. It follows from Theorem 6.1 in particular that

$$\lim_{n\to\infty} (a*u_n - M_a^{(0)}u_n) = 0$$

if $M_a^{(0)}$ is finite.

In the positive recurrent case $M_b = M_B^{(0)}$ is finite, and we can choose $a = B$ to get

$$\lim_{n\to\infty} (1 - M_b u_n) = 0.$$

Thus we have proved the result in the case where M_b and $M_a^{(0)}$ are finite. In the case $M_b < \infty$, $M_a^{(0)} = \infty$, the result easily follows by using a truncation argument.

So it remains to consider the null recurrent case; this is the case where $M_b = \infty$. Let $\varepsilon > 0$ be arbitrary and $N = N(\varepsilon) \geq 0$ be such that

$$\sum_0^N B_n \geq \varepsilon^{-1}.$$

Since by (6.4),

$$\lim_{n\to\infty} (u_n - u_{n-1}) = 0,$$

we have

$$\min_{0\leq m\leq N} u(n-m) \geq u(n) - \varepsilon$$

for n big enough. By (4.6)

$$1 \geq \sum_{m=0}^{N} B(m)u(n-m) \geq \varepsilon^{-1}(u(n) - \varepsilon)$$

for large n implying

$$\lim_{n\to\infty} u_n = 0.$$

By suitable truncation the final result follows:

$$\lim_{n\to\infty} a*u_n = 0. \qquad \square$$

Let now $u = (u_n; n \geq 0)$ be an arbitrary (possibly non-probabilistic) renewal sequence with convergence parameter $0 < R < \infty$. Applying Theorems 6.1 and 6.2 to the renewal sequence $\tilde{u} = (R^n u_n; n \geq 0)$ (see (4.8)) gives us the following corollary. We write $\tilde{B}_n = 1 - \sum_1^n R^m b_m (= \sum_{n+1}^{\infty} R^m b_m$, whenever u is R-recurrent).

Corollary 6.1. Let $u = (u_n; n \geq 0)$ be an aperiodic, R-recurrent renewal sequence. Then:

(i) For any non-negative sequence $a = (a_n; n \geq 0)$ such that $\hat{a}(R) = \sum_0^{\infty} R^n a_n$ is finite,

$$\lim_{n \to \infty} \sum_{m=0}^{n} R^{n-m} |a * u_{n-m} - \hat{a}(R) u_{n-m}| \tilde{B}_m = 0.$$

(ii) There exists $\tilde{u}_{\infty} = \lim_{n \to \infty} R^n u_n = M_{\tilde{b}}^{-1}$. Moreover $\tilde{u}_{\infty} > 0$ or $= 0$ depending on whether the renewal sequence u is R-positive or R-null recurrent. □

Let us return to the probabilistic case. We assume that $(T(i); i \geq 0)$ is an aperiodic, recurrent renewal process. $u = (u_n; n \geq 0)$ is the corresponding undelayed renewal sequence. If we want to emphasize a specific delay distribution $a = (a_n; n \geq 0)$, we write $\mathbb{P}_a$ instead of $\mathbb{P}$ for the underlying probability measure, i.e. $\mathbb{P}_a\{T(0) = n\} = a_n, n \geq 0$.

We shall study the asymptotic behaviour of the sums

$$\sum_0^N a * u_n = a * u * 1_N$$

as $N \to \infty$. In probabilistic terms the above sum is equal to the expectation of the number of renewals in the interval $[0, N]$ if the delay distribution is a,

$$\sum_0^N a * u_n = \mathbb{E}_a \sum_0^N 1_{\{Y_n = 1\}};$$

by recurrence it tends to ∞ as $N \to \infty$. We denote

$$M_a^{(2)} = \sum_0^{\infty} n^2 a_n,$$

$$A_n = \sum_{n+1}^{\infty} a_m = 1 - a * 1_n, \quad n \geq 0.$$

Theorem 6.3. With the assumptions of Theorem 6.1:

(i) We have

$$\lim_{N \to \infty} \sum_0^N (a * u_n - u_n) = -M_b^{-1} M_a, \tag{6.5}$$

unless $M_a = M_b = \infty$.

(ii) If M_b is finite, then

$$\lim_{N\to\infty}\left[\sum_0^N a*u_n-(N+1)M_b^{-1}\right]$$

$$=\tfrac{1}{2}M_b^{-2}(M_b^{(2)}-M_b)-M_b^{-1}M_a, \tag{6.6}$$

unless $M_a=M_b^{(2)}=\infty$, and in particular,

$$\lim_{N\to\infty}\left[\sum_0^N u_n-(N+1)M_b^{-1}\right]=\tfrac{1}{2}M_b^{-2}(M_b^{(2)}-M_b). \tag{6.7}$$

Proof. This theorem is in fact a corollary of Theorem 6.2. Namely, we can write the term on the left hand side of (6.5) in the form

$$\sum_0^N(a*u_n-u_n)=a*u*1_N-u*1_N=-A*u_n.$$

By Theorem 6.2 $A*u_n$ tends to the limit

$$M_b^{-1}M_A^{(0)}=M_b^{-1}M_a \quad \text{as } N\to\infty,$$

unless $M_b=\infty$ and $M_A^{(0)}=M_a=\infty$.

Setting a to be equal to the equilibrium distribution e,

$$a=e=M_b^{-1}B,$$

and using (4.6) we obtain

$$\lim_{N\to\infty}\left(\sum_0^N u_n-(N+1)M_b^{-1}\right)=\lim_{N\to\infty}\sum_0^N(u_n-e*u_n)=M_b^{-1}M_e.$$

Now it is easy to see that

$$M_e=\tfrac{1}{2}M_b^{-1}(M_b^{(2)}-M_b).$$

This proves (6.7). Clearly, (6.6) is a direct consequence of (6.5) and (6.7). □

According to Theorem 6.3 the series

$$\sum_0^\infty(a*u_n-u_n)$$

converges provided that $M_a<\infty$, and the series

$$\sum_0^\infty(u_n-M_b^{-1})$$

converges provided that $M_b^{(2)}<\infty$. The following theorem deals with the absolute convergence of these series.

Theorem 6.4. With the assumptions of Theorem 6.1:

(i) If M_b is finite, then the *total variation* Var (u) of the renewal sequence u,

$$\operatorname{Var}(u) \overset{\text{def}}{=} 1 + \sum_{1}^{\infty} |u_n - u_{n-1}|,$$

is finite. If in addition M_a is finite, then

$$\sum_{0}^{\infty} |a * u_n - u_n| \leq M_a \operatorname{Var}(u) < \infty. \tag{6.8}$$

(ii) If $M_b^{(2)}$ is finite, then

$$\sum_{0}^{\infty} |u_n - M_b^{-1}| \leq \tfrac{1}{2} M_b^{-1}(M_b^{(2)} - M_b) \operatorname{Var}(u) < \infty, \tag{6.9}$$

and

$$\operatorname{Var}^{(2)}(u) \overset{\text{def}}{=} \sum_{1}^{\infty} n |u_n - u_{n-1}| < \infty. \tag{6.10}$$

If in addition, $M_a^{(2)}$ is finite, then

$$\sum_{1}^{\infty} n |a * u_n - u_n| \leq \tfrac{1}{2}(M_a^{(2)} - M_a) \operatorname{Var}(u) + M_a \operatorname{Var}^{(2)}(u) < \infty. \tag{6.11}$$

Proof. The proof of the finiteness of the total variation Var (u) is based again on the use of the coupling technique.

Let $T = (T(i); i \geq 0)$ and $T' = (T'(i); i \geq 0)$ be two *independent* renewal processes, the former being undelayed, $T(0) = 0$, the latter being delayed with delay $T'(0) = 1$, and both having the same increment distribution b. Consider the *product renewal process* $T'' = (T''(i); i \geq 0)$, that is, the renewal process with incidence process $Y''_n = Y_n Y'_n, n \geq 0$. Thus the renewal epochs of T'' consist of precisely the common renewal epochs of the renewal processes T and T'. It follows that the delayed renewal sequence v'' corresponding to T'' is given by

$$v''_0 = 0, \quad v''_n = u_n u_{n-1} \quad \text{for } n \geq 1.$$

By Theorem 6.2,

$$\lim_{n \to \infty} v''_n = u_\infty^2 = M_b^{-2} > 0.$$

Consequently, the product renewal process T'' is positive recurrent. This implies that the bivariate backward chain (V_n, V'_n) is positive recurrent. (To see this, note that $Y''_n = 1$ if and only if $(V_n, V'_n) = (0, 0)$.)

Since by Theorem 6.2

$$u_n \geq (2M_b)^{-1} > 0 \quad \text{for all sufficiently big } n,$$

we can easily prove that the bivariate chain (V_n, V'_n) is aperiodic and irreducible positive recurrent on the state space $E_b \times E_b$, where $E_b = \{0, 1, \ldots, \bar{M}-1\}$ or $E_b = \mathbb{N}$ depending on whether

$$\bar{M} = \sup\{n \geq 1 : b(n) > 0\}$$

is finite or infinite. Since $T''(0) = S_{(0,0)}$ is the first time at which the bivariate chain (V_n, V'_n) hits state $(0,0)$, the expectation $\mathbb{E}[T''(0)|(V_0, V'_0) = (1,0)] = \mathbb{E}_{(1,0)}[S_{(0,0)}]$ is finite by Corollary 5.4. $\mathbb{E}T''(0)$ is finite, since clearly,

$$\mathbb{E}T''(0) \leq 1 + \mathbb{E}[T''(0)|(V_0, V'_0) = (1,0)].$$

Using precisely the same arguments as in the proof of Theorem 6.1 we obtain the coupling inequality

$$|u_n - u_{n-1}| \leq \mathbb{P}\{T''(0) > n\} \quad \text{for all } n \geq 1, \tag{6.12}$$

from which the finiteness of Var (u) follows after summing n over $\mathbb{N}$. In order to obtain (6.8) set $w_0 = 1, w_n = |u_n - u_{n-1}|$ for $n \geq 1$, and note that

$$\sum_{n=0}^{\infty} |a * u_n - u_n| \leq A * w_n. \tag{6.13}$$

The assertion now follows, since $\sum_0^\infty A_n = M_a$ and $\sum_0^\infty w_n = \text{Var}(u)$.

(ii) The inequalities (6.9) follow from (i) by setting $a = e$. In order to prove (6.10) we adopt the notation used in the proof of part (i), and note first that by Proposition 5.18 $\mathbb{E}_{(1,0)}[S^2_{(0,0)}]$ is finite and hence so is $\mathbb{E}[T''(0)^2]$. The assertion (6.10) follows now easily from the coupling inequality (6.12).

The inequality (6.11) can be proved by a straightforward calculation. □

Let $\mathscr{A}$ be any family of delay distributions a such that

$$\lim_{n\to\infty} A_n = \lim_{n\to\infty} \sum_{n+1}^{\infty} a_m = 0 \quad \text{uniformly over } a \in \mathscr{A}.$$

Note that a sufficient condition for this is:

$$\sup_{a \in \mathscr{A}} \mathbb{E}_a T(0) = \sup_{a \in \mathscr{A}} \sum_0^{\infty} n a_n < \infty.$$

It turns out that uniform convergence of the tails of the initial distributions leads to uniform convergence in the renewal theorem:

Theorem 6.5. With the assumptions of Theorem 6.1: If M_b is finite and $\lim_{n\to\infty} A_n = 0$ uniformly over $a \in \mathscr{A}$, then

$$\lim_{n\to\infty} a * u_n = M_b^{-1} \quad \text{uniformly over } a \in \mathscr{A}.$$

Proof. By (6.13) $|a * u_n - u_n|$ is dominated by $A * w_n$. This is further dominated by

$$\sum_{m=n-N+1}^{\infty} w_m + A_N \operatorname{Var}(u), \quad \text{for all } 0 \le N \le n.$$

By hypothesis and Theorem 6.4, the first term on the right hand side tends to zero as $n \to \infty$. By hypothesis, the second term tends to zero uniformly over $\mathscr{A}$ as $N \to \infty$. □

Our next aim is to investigate the case where the rate of convergence in the renewal theorem is geometric. If the renewal sequence $(u_n; n \ge 0)$ tends to its limit $u_\infty = M_b^{-1}$ with a *geometric rate*, i.e. for some constants $M < \infty$ and $\rho < 1$,

$$|u_n - u_\infty| = |u_n - M_b^{-1}| \le M\rho^n \quad \text{for all } n \ge 0,$$

then the renewal process $(T(i); i \ge 0)$ is called *geometrically ergodic*. In the following theorem necessary and sufficient conditions are given for the geometric ergodicity of T.

Theorem 6.6. With the assumptions of Theorem 6.1 the following three conditions are equivalent:

(i) The renewal process $(T(i); i \ge 0)$ is geometrically ergodic.

(ii) The series

$$\hat{b}(r) = \sum_{1}^{\infty} r^n b_n$$

converges for some $r > 1$.

(iii) There is a constant $r_0 > 1$ such that the function $\hat{u}$ defined on the complex plane $\mathbb{C}$ by

$$\hat{u}(z) = \sum_{0}^{\infty} z^n u_n$$

has no singularities in the disc $\{|z| < r_0\}$ except a simple pole at $z = 1$.

Proof. Assume first that (i) holds. Denote $f_0 = 1, f_n = u_n - u_{n-1}$ for $n \ge 1$. Then the function

$$\hat{f}(z) = \sum_{0}^{\infty} z^n f_n = (1 - z)\hat{u}(z)$$

has no singularities in the disc $\{|z| < \rho^{-1}\}$, i.e., we have (iii).

Conversely, if (iii) holds, then the function $(1 - z)\hat{u}(z) = \hat{f}(z)$ is regular in the disc $\{|z| < r_0\}$ implying

$$\sum_{0}^{\infty} r^n |u_n - u_{n-1}| < \infty \quad \text{for all } r < r_0.$$

This clearly leads to (i).

By the renewal equation (4.3)

$$\hat{f}(z) = (1-z)\hat{u}(z) = (1-z)(1-\hat{b}(z))^{-1} \quad \text{in the disc } \{|z|<1\}. \tag{6.14}$$

If (ii) holds, then $\hat{b}(z)$ is regular in the disc $\{|z|<r\}$. Consequently, the equation

$$\hat{b}(z) = 1$$

has only finitely many roots in the disc $\{|z| \le r-\delta\}, \delta > 0$ arbitrary. Since, by aperiodicity, there is only one root, namely $z=1$ in the disc $\{|z|\le 1\}$, it follows that there exists $r_0 > 1$ such that $z=1$ is the only root in the disc $\{|z|<r_0\}$. By (6.14), we have (iii). Also the converse implication (iii)$\Rightarrow$(ii) follows easily from (6.14). □

6.2 Convergence of the iterates $K^n(x, A)$

In the following we shall study the convergence of the iterates $K^n(x, A)$ as $n \to \infty$. This will now be relatively easy, since $K^n(x, A), n \ge 0$, is by the decomposition results of Theorem 4.1 essentially equal to a delayed renewal sequence, and so the renewal theorems of the previous section can be directly applied to it.

Throughout Sections 6.2–6.7 *we assume that K is an R-recurrent kernel.* We adopt the earlier notation. In particular, d denotes the period of K, and $m_0 \ge 1, s \in \mathscr{S}^+$ and $\nu \in \mathscr{M}^+$ denote fixed quantities such that the minorization condition $M(m_0, 1, s, \nu)$ holds. We assume that $c_{m_0} = \text{g.c.d.}\{m_0, d\} = 1$ (cf. (5.1)). π denotes a fixed R-invariant measure and $h = h_\nu = R^{m_0} G^{(R)}_{m_0,s,\nu} s = \sum_0^\infty R^{(n+1)m_0}(K^{m_0} - s \otimes \nu)^n s$ is the unique minimal R-invariant function satisfying $\nu(h) = 1$ (cf. Theorem 5.1). *If K is R-positive recurrent, then we shall, by convention, normalize π so that $\pi(h) = 1$.* The transition probability $\tilde{K} = RI_{h^{-1}}KI_h$ governs the transitions of a Harris recurrent Markov chain on the closed set $\tilde{E} = \{h < \infty\}$ (cf. Proposition 5.4).

For any $g \in \mathscr{E}_+$, we write $\mathscr{M}(g) = \{\lambda \in \mathscr{M} : g\ |\lambda|\text{-integrable}\}$. The space $\mathscr{M}(g)$ is equipped with the *g-total variation norm*: for a signed measure $\lambda \in \mathscr{M}(g)$,

$$\|\lambda\|_g \overset{\text{def}}{=} \sup_{|f| \le g} |\lambda(f)| = |\lambda|(g).$$

$\|K\|_g$ denotes the corresponding operator norm:

$$\|K\|_g = \sup_{\|\lambda\|_g \le 1} \|\lambda K\|_g = \sup_{x \in \{g<\infty\}} g(x)^{-1} Kg(x).$$

When $g \equiv 1$, we have $\mathscr{M}(g) = \mathscr{M}(1) = \text{b}\mathscr{M}$. Then $\|\lambda\|_g = \|\lambda\|_1$ is simply the *total variation* of the bounded signed measure λ:

$$\|\lambda\|_1 = \|\lambda\| \overset{\text{def}}{=} \sup_{A\in\mathscr{E}} \lambda(A) - \inf_{A\in\mathscr{E}} \lambda(A) = |\lambda|(E).$$

For a bounded kernel $\|K\|_1 = \sup_{x\in E} K(x, E) < \infty$. For a substochastic kernel P, $\|P\|_1 \leq 1$; i.e., P is a *contraction* on b$\mathscr{M}$.

Proposition 6.1. The operator norm $\|K\|_h$ equals R^{-1}.

Proof. By R-invariance $\|K\|_h = \sup_{x\in\{h<\infty\}} h(x)^{-1}Kh(x) = R^{-1}$. □

For any $g\in\mathscr{E}_+$, we write

$$\mathscr{M}_0(g) = \{\lambda\in\mathscr{M}(g): \lambda(g) = 0\};$$

i.e., a signed measure $\lambda\in\mathscr{M}$ is a member of $\mathscr{M}_0(g)$ whenever g is $|\lambda|$-integrable and the integral $\lambda(g)$ vanishes.

We start by proving an h-total variation convergence result for aperiodic K. We will discuss the periodic case later in Corollary 6.6.

Theorem 6.7. Suppose that K is an aperiodic, R-recurrent kernel. Then for any signed measure $\lambda\in\mathscr{M}_0(h)$:

$$\lim_{n\to\infty} R^n \|\lambda K^n\|_h = 0.$$

Proof. It is no restriction to consider the case $R = 1$ only. (Otherwise we should look at the 1-recurrent kernel RK.)

Assume first that $m_0 = 1$, i.e., the pair (s, ν) is an atom. By the decomposition result (iv) of Theorem 4.1

$$\|\lambda K^n\|_h \leq |\lambda|(K - s\otimes\nu)^n h + |\lambda(a) * u| * \sigma(h)_{n-1} \text{ for all } n \geq 1. \tag{6.15}$$

(We write $\lambda(a) = (\lambda(a_n); n \geq 0)$; the notation $\sigma(h)$ is interpreted similarly.) For the first term on the right hand side we have the estimates

$$\begin{aligned} \infty > |\lambda|(h) &\geq |\lambda|(K - s\otimes\nu)^n h \\ &= |\lambda| \sum_{m=n}^{\infty} (K - s\otimes\nu)^m s \downarrow 0 \quad \text{as } n\to\infty. \end{aligned}$$

In order to see that the second term also tends to zero as $n\to\infty$, observe that, by (4.11)

$$\sum_0^\infty |\lambda(a_n)| \leq |\lambda|\left(\sum_0^\infty a_n\right) = |\lambda|(h) < \infty,$$

$$\sum_0^\infty \lambda(a_n) = \lambda\left(\sum_0^\infty a_n\right) = \lambda(h) = 0$$

and by (4.22)

$$\sigma_n(h) = B_n,$$

and then use Theorem 6.1.

The general case, $m_0 \geq 1$, follows by considering the iterated kernel K^{m_0} and using the fact that $\| K \|_h = 1$ (see Proposition 6.1). □

Setting $\lambda = h(x)^{-1}\varepsilon_x - h(y)^{-1}\varepsilon_y$ in Theorem 6.7, where $x, y \in \tilde{E}$, we get:

Corollary 6.2. Suppose that K is aperiodic R-recurrent. Then for any states $x, y \in \tilde{E}$:

$$\lim_{n \to \infty} R^n \| h(x)^{-1} K^n(x, \cdot) - h(y)^{-1} K^n(y, \cdot) \|_h = 0. \quad \square$$

Setting $\lambda = \mu - \mu(h)\pi$, where μ belongs to $\mathcal{M}(h)$, we get the following corollary. (By convention $\pi(h) = 1$, if K is R-positive recurrent.)

Corollary 6.3. Suppose that K is aperiodic R-positive recurrent. Then for any signed measure $\mu \in \mathcal{M}(h)$:

$$\lim_{n \to \infty} \| R^n \mu K^n - \mu(h)\pi \|_h = 0. \quad \square$$

For every $n \geq 1$, let $k^{(n)} \in (\mathcal{E} \otimes \mathcal{E})_+$ and $K_s^{(n)}$ denote the density and singular part, respectively, of K^n w.r.t. π (cf. Lemma 2.6):

$$K^n(x, \mathrm{d}y) = k^{(n)}(x, y)\pi(\mathrm{d}y) + K_s^{(n)}(x, \mathrm{d}y).$$

Since for all $x \in \tilde{E}$, $n \geq 0$.

$$\| R^n K^n(x, \cdot) - h(x)\pi \|_h$$
$$= \int |R^n k^{(n)}(x, y) - h(x)| h(y)\pi(\mathrm{d}y) + R^n K_s^{(n)} h(x),$$

we obtain from Corollary 6.3:

Corollary 6.4. Suppose that K is aperiodic R-positive recurrent. Then for all $x \in \tilde{E}$:

$$\lim_{n \to \infty} \int |R^n k^{(n)}(x, y) - h(x)| h(y)\pi(\mathrm{d}y) = 0$$

and

$$\lim_{n \to \infty} R^n K_s^{(n)} h(x) = 0. \quad \square$$

It follows in particular from Corollary 6.3 that, whenever K is aperiodic

R-positive recurrent, then, for all $x \in \tilde{E} = \{h < \infty\}$, and all $f \in \mathscr{E}_+$ satisfying $0 \le f \le Mh$ for some constant $M < \infty$, the limit

$$\lim_{n \to \infty} R^n K^n f(x) = h(x)\pi(f) \tag{6.16}$$

exists (and is finite). It turns out that (6.16) holds even for $x \in (\tilde{E})^c = \{h = \infty\}$:

Proposition 6.2. Suppose that K is aperiodic R-positive recurrent. Then for all $x \in E$ such that $h(x) = \infty$, for all $f \in \mathscr{E}^+$:

$$\lim_{n \to \infty} R^n K^n f(x) = \infty.$$

Proof. Use Theorem 4.1(iv) (cf. the proof of Theorem 6.7). □

The following result is the 'dual' of Theorem 6.7. First recall the definition of the $\mathscr{L}^1(\pi)$-norms:

$$\| f \|_\pi = \pi(|f|), \quad f \in \mathscr{L}^1(\pi).$$

We set $\mathscr{L}^1_0(\pi) = \{f \in \mathscr{L}^1(\pi) \colon \pi(f) = 0\}$.

Theorem 6.8. Suppose that K is aperiodic R-recurrent. Then for any function $f \in \mathscr{L}^1_0(\pi)$

$$\lim_{n \to \infty} R^n \| K^n f \|_\pi = 0.$$

Proof. The proof is analogous to that of Theorem 6.7. First note again that we can assume $R = 1$ and $m_0 = 1$. The inequality corresponding to the inequality (6.15) is

$$\| K^n f \|_\pi \le \pi(K - s \otimes \nu)^n |f| + \pi(a) * |u * \sigma(f)|_{n-1}. \quad \square$$

Corollary 6.5. Suppose that K is aperiodic R-positive recurrent. Then for any $f \in \mathscr{L}^1(\pi)$:

$$\lim_{n \to \infty} \| R^n K^n f - \pi(f) h \|_\pi = 0. \quad \square$$

It turns out that the convergence in Corollary 6.5 is also pointwise (π-a.e.):

Theorem 6.9. Suppose that K is aperiodic R-recurrent. Then for any $f \in \mathscr{L}^1(\pi)$, π-almost all $x \in E$:

$$\lim_{n \to \infty} R^n K^n f(x) = h(x)\pi(h)^{-1}\pi(f).$$

Proof. There is no loss of generality in assuming that $K = P$ is a Harris recurrent transition probability. (Otherwise we should make the similarity transform (5.2); see also Proposition 5.4.) Also we can assume that $m_0 = 1$.

Let now $x \in R_{|f|}$ ($\stackrel{\text{def}}{=}$ the set of $|f|$-regular states $\stackrel{\pi\text{-a.e.}}{=} E$, see Proposition 5.13(ii)) be arbitrary. We have

$$P^n f(x) = (P - s \otimes \nu)^n f(x) + a(x) * u * \sigma(f)_{n-1}.$$

Since by Proposition 5.13(ii) the sum $\sum_0^\infty (P - s \otimes \nu)^n f(x) = G_{s,\nu} f(x)$ is finite, the first term on the right hand side tends to zero. The latter clearly tends to $\pi(E)^{-1}\pi(f)$. □

Note that the limit in the above theorem is equal to zero (for π-almost all $x \in E$) whenever K is R-null recurrent.

If K is R-null recurrent, then $R^n \lambda K^n f$ tends to zero, for all $\lambda \in \mathcal{M}(h)$ and $f \in \mathcal{L}^1(\pi)$ with $|f| \le Mh$ for some constant $M < \infty$. In fact, slightly more holds true:

Theorem 6.10. Suppose that K is aperiodic R-null recurrent. Then for any $\lambda \in \mathcal{M}_+(h)$ and any constant $\gamma > 0$:

$$\lim_{n\to\infty} \frac{R^n \lambda K^n f}{\pi(f) + \gamma} = 0 \quad \text{uniformly over the class } \{f \in \mathcal{E}_+ : f \le h\}.$$

Proof. There is no loss of generality in assuming that $\lambda(h) = 1$. Let $\varepsilon, \gamma > 0$ be arbitrary. By Theorem 6.7 and Egoroff's theorem, and since K is R-null recurrent, there is a set $B = B(\varepsilon) \in \mathcal{E}$ with $\int_B \pi(\mathrm{d}x) h(x) \ge \varepsilon^{-1}$ and an integer $N = N(\varepsilon)$ such that

$$R^n \| h(x)^{-1} K^n(x, \cdot) - \lambda K^n \|_h \le \gamma\varepsilon \quad \text{for all } x \in B,\ n \ge N.$$

Consequently, we have for all $0 \le f \le h$, $n \ge N$:

$$\begin{aligned} \pi(f) &= R^n \pi K^n f \\ &\ge R^n \int_B \pi(\mathrm{d}x) K^n f(x) \\ &\ge \int_B \pi(\mathrm{d}x) h(x) (R^n \lambda K^n f - \gamma\varepsilon) \\ &\ge \varepsilon^{-1} R^n \lambda K^n f - \gamma, \end{aligned}$$

from which the result follows. □

We shall now consider the case where K has period $d \ge 2$. Recall from Proposition 5.5 that, for each $i = 0, \ldots, d-1$, the function h restricted to the cyclic set E_i is the (essentially unique) minimal R^d-invariant function for the R^d-recurrent, aperiodic kernel K^d with state space E_i. Hence we have:

Corollary 6.6. Suppose that K is R-recurrent and periodic with period

$d \geq 2$ and cyclic sets $E_0, \ldots, E_{d-1}$. Let $N = (E_0 + \cdots + E_{d-1})^c$. Then for any $\lambda \in \mathcal{M}_0(h)$ such that $\int_{E_i} \lambda(dx)h(x) = 0$ for each $i = 0, \ldots, d-1$, and $|\lambda|(N) = 0$:

$$\lim_{n \to \infty} R^n \| \lambda K^n \|_h = 0. \qquad \square$$

In the important special case, where $K = P$ is the transition probability of a Harris recurrent Markov chain (X_n), we have $h \equiv 1$, and consequently we obtain from the preceding general results:

Corollary 6.7. Suppose that (X_n) is a Harris recurrent Markov chain. Let λ and μ be any two initial distributions and let $f, g \in \mathcal{L}^1(\pi)$ be such that $\pi(f) = \pi(g)$. Then:

(i) If (X_n) is aperiodic, then

$$\lim_{n \to \infty} \| \lambda P^n - \mu P^n \| = 0, \tag{6.17}$$

$$\lim_{n \to \infty} \| P^n f - P^n g \|_\pi = 0,$$

and

$$\lim_{n \to \infty} P^n f(x) = \frac{\pi(f)}{\pi(E)} \quad \text{for } \pi\text{-almost all } x \in E.$$

(ii) If (X_n) is aperiodic and positive Harris recurrent with stationary distribution π, then

$$\lim_{n \to \infty} \| \lambda P^n - \pi \| = 0 \tag{6.18}$$

and

$$\lim_{n \to \infty} \| P^n f - \pi(f) 1 \|_\pi = 0;$$

in particular, writing respectively $p^{(n)} \in (\mathscr{E} \otimes \mathscr{E})_+$ and $P_s^{(n)}$ for the absolutely continuous part and the singular part of P^n w.r.t. π, we have for all $x \in E$:

$$\lim_{n \to \infty} \| p^{(n)}(x, \cdot) - 1 \|_\pi = 0$$

and

$$\lim_{n \to \infty} P_s^{(n)}(x, E) = 0.$$

(iii) If (X_n) is aperiodic and null recurrent then for any initial distribution λ and any constant $\gamma > 0$:

$$\lim_{n \to \infty} \sup_{A \in \mathscr{E}} \frac{\lambda P^n(A)}{\pi(A) + \gamma} = 0.$$

(iv) If (X_n) is periodic with period $d \geq 2$ and cyclic sets $E_0, \ldots, E_{d-1}$, and if $\lambda(E_i) = \mu(E_i)$ for all $i = 0, \ldots, d-1$, and $\lambda(N) = \mu(N) = 0$ $(N = (E_0 + \cdots + E_{d-1})^c)$, then

$$\lim_{n \to \infty} \| \lambda P^n - \mu P^n \| = 0. \qquad \square$$

Examples 6.1. (*a*) Suppose that $K = (k(x,y);\ x,y \in E)$ is a Card_F-irreducible, aperiodic, R-recurrent matrix on a discrete state space E. Let $h = (h(x); x \in E)$ denote the (essentially unique) R-invariant column vector for K. Then for any two states $x, y \in F$:

$$\lim_{n \to \infty} R^n \sum_{z \in E} \left| h(x)^{-1} k^n(x,z) - h(y)^{-1} k^n(y,z) \right| h(z) = 0.$$

(*e*) Consider the renewal process $(Z_n; n \geq 0)$ on $\mathbb{R}_+$ introduced in Example 1.2(*e*). Let us write u for the corresponding *renewal measure*,

$$u(A) \overset{\text{def}}{=} \sum_0^\infty F^{*n}(A) = \mathbb{E}\left[\sum_0^\infty 1_A(Z_n) \middle| Z_0 = 0 \right], \quad A \in \mathscr{R}_+.$$

Let $B(t) = 1 - F(t) = \mathbb{P}\{z_1 > t\}$. Then $u * B \equiv 1$.

Suppose that the increment distribution F is spread out. Then for any bounded interval $[0, c]$, $0 < c < \infty$, and constant $0 < \gamma < \infty$:

$$\lim_{t \to \infty} \sup_{A \in \mathscr{R}_+ \cap [0,c]} \left| u(t + A) - u(t - \gamma + A) \right| = 0.$$

(Hints for the proof: for all $t \geq \gamma > 0$, $A \in \mathscr{R}_+$:

$$u(t - \gamma + A) = P^t(\gamma, \cdot) * u(A),$$

where P^t denotes the transition probability of the forward process, $P^t(x, \mathrm{d}y) = \mathbb{P}\{V_t^+ \in \mathrm{d}y | V_0^+ = x\}$.)

6.3 Ergodic Markov chains

We shall now for a while consider the case where $K = P$ is the transition probability of a Markov chain (X_n). We shall resume the general K in Section 6.7.

We call a Markov chain (X_n) (Harris) *ergodic*, if it is aperiodic and positive Harris recurrent. By Corollary 6.7(ii),

$$\lim_{n \to \infty} \| \lambda P^n - \pi \| = 0$$

for any initial distribution λ. There is, in fact, also a converse result to that:

Proposition 6.3. A Markov chain (X_n) is ergodic if (and only if) there exists a probability measure π on $(E, \mathscr{E})$ such that

$$\lim_{n \to \infty} \| P^n(x, \cdot) - \pi \| = 0 \quad \text{for all } x \in E. \tag{6.19}$$

Proof. Suppose that (6.19) holds. Clearly, (X_n) is π-irreducible and aperiodic. Moreover, it follows that $G(x, A) = \infty$ for all $x \in E$, π-positive $A \in \mathscr{E}$. Thus (X_n) is recurrent. By Corollary 6.7(iii) (X_n) is positive recurrent. Let h be an arbitrary bounded harmonic function for (X_n). From (6.19) it follows that $h \equiv \pi(h)$, whence, by Theorem 3.8(ii) (X_n) is Harris recurrent. □

Throughout Sections 6.3–6.6 *we shall assume that* $K = P$ *is the transition probability of an ergodic Markov chain* (X_n). Our aim in these sections is to study the rate of convergence in (6.17) and (6.18). It turns out that these rates are closely connected with the degrees of recurrence of (X_n).

We shall not formulate the results in the periodic case. Corollary 6.6 provides an example of how this extension could be performed. The same remark holds for the formulation of the dual results (cf. Theorem 6.8).

Let us fix two initial distributions λ and μ of (X_n). Recall from Definition 5.4 the concept of regularity: The probability measure λ is regular, if $\mathbb{E}_\lambda S_B$ is finite for all $B \in \mathscr{E}^+$. We shall show that, for λ and μ regular, the rate of convergence in (6.17) is n^{-1}.

For the following theorem note that, since the σ-algebra $\mathscr{E}$ is countably generated, the function $(x, y) \to \| P^n(x, \cdot) - P^n(y, \cdot) \|$ is measurable for all $n \geq 1$.

Also note that, for any non-negative non-increasing summable sequence $(c_n; n \geq 0)$, $\lim_{n \to \infty} nc_n = 0$.

Theorem 6.11. Suppose that (X_n) is ergodic. If λ and μ are regular then

$$\sum_{n=0}^{\infty} \iint \lambda(\mathrm{d}x)\mu(\mathrm{d}y) \| P^n(x, \cdot) - P^n(y, \cdot) \| < \infty.$$

Proof. Consider first the case where $m_0 = 1$, i.e., (s, v) is an atom for P. By Theorem 4.1, (4.23) and (4.24) we have for all $n \geq 1$, $x \in E$:

$$\| P^n(x, \cdot) - vP^{n-1} \| \leq \mathbb{P}_x\{T_\alpha \geq n\} + |a(x) * u - u| * \sigma(E)_{n-1}. \quad (6.20)$$

By Theorem 6.4(i), and by (4.21) and (4.22),

$$\sum_{n=0}^{\infty} |a(x) * u_n - u_n| \leq \mathbb{E}_x[T_\alpha] \operatorname{Var}(u),$$

$$\sum_{n=0}^{\infty} \sigma_n(E) = \mathbb{E}_\alpha S_\alpha = \pi(s)^{-1}.$$

Consequently,

$$\sum_{n=1}^{\infty} \int \lambda(\mathrm{d}x) \| P^n(x, \cdot) - vP^{n-1} \| \leq \mathbb{E}_\lambda T_\alpha + \pi(s)^{-1} \mathbb{E}_\lambda T_\alpha \operatorname{Var}(u),$$

which is finite by Proposition 5.13(iv) and Theorem 6.4(i). The final result in the case $m_0 = 1$ now follows by using the triangle inequality.

To prove the assertion for general m_0 use Lemma 5.3 and the contractivity of P. □

From the preceding theorem and from Proposition 5.13(ii) we obtain the following:

Corollary 6.8. Suppose that (X_n) is ergodic.

(i) If λ and μ are regular then

$$\sum_0^\infty \| \lambda P^n - \mu P^n \| < \infty;$$

in particular,

$$\lim_{n\to\infty} n \| \lambda P^n - \mu P^n \| = 0.$$

(ii) For π-almost all $x, y \in E$:

$$\sum_{n=0}^\infty \| P^n(x,\cdot) - P^n(y,\cdot) \| < \infty$$

and

$$\lim_{n\to\infty} n \| P^n(x,\cdot) - P^n(y,\cdot) \| = 0. \quad \square$$

According to the above corollary we know that, if λ and μ are regular, then the sums

$$\mathbb{E}_\lambda \sum_0^N f(X_n) - \mathbb{E}_\mu \sum_0^N f(X_n) = \sum_0^N (\lambda P^n f - \mu P^n f)$$

converge (even uniformly over $-1 \le f \le 1$) as $N \to \infty$. This naturally raises the problem of identifying the limit. For the following theorem recall the definition of the kernel

$$\bar{G}_{m_0,s,\nu} = \sum_{n=0}^\infty (P^{m_0} - s \otimes \nu)(I + \cdots + P^{m_0 - 1}).$$

Theorem 6.12. Suppose that (X_n) is ergodic.

If λ is regular then for all $f \in b\mathscr{E}$:

$$\sum_0^\infty (\lambda P^n f - \nu P^n f) = \lambda \bar{G}_{m_0,s,\nu}(I - 1 \otimes \pi) f.$$

In particular, there is a full set $F(= R_1)$ such that for all $x, y \in F$, all $f \in b\mathscr{E}$:

$$\sum_0^\infty (P^n f(x) - \nu P^n f) = \bar{G}_{m_0,s,\nu}(I - 1 \otimes \pi) f(x).$$

Proof. Assume first that $m_0 = 1$. By Theorem 4.1, Theorems 6.3 and 6.4, and since $M_b = \pi(s)^{-1}$, $M_{\lambda(a)} = \mathbb{E}_\lambda T_\alpha = \lambda G_\alpha 1 - 1$, we have

$$\sum_0^N \lambda P^n f - \sum_1^{N+1} \nu P^{n-1} f$$

$$= \mathbb{E}_\lambda \sum_0^{T_\alpha \wedge N} f(X_n) + \sum_1^N (\lambda(a) * u - u) * \sigma(f)_{n-1} - \nu P^N f$$

$$\underset{N\to\infty}{\to} \mathbb{E}_\lambda \sum_0^{T_\alpha} f(X_n) + \sum_{m=0}^{\infty} (\lambda(a) * u_m - u_m) \sum_{n=0}^{\infty} \sigma_n(f) - \pi(f)$$

$$= \lambda G_\alpha f - M_b^{-1} M_{\lambda(a)} \pi(s)^{-1} \pi(f) - \pi(f)$$

$$= \lambda G_\alpha f - \lambda G_\alpha 1 \pi(f).$$

The general case, $m_0 \geq 1$, can easily be proved by considering the m_0-step chain (X_{nm_0}) and the function $f + \cdots + P^{m_0 - 1} f$ instead of f. □

Examples 6.2. (*a*) Suppose that $P = (p(x, y); x, y \in E)$ is an ergodic transition matrix with stationary distribution $\pi = (\pi(x); x \in E)$. Set $E_\pi = \{\pi > 0\}$ (cf. Example 5.1(*a*)). Then for all $x \in E$,

$$\lim_{n\to\infty} \sum_{z \in E} |p^n(x, z) - \pi(z)| = 0,$$

and for all $x, y \in E_\pi$,

$$\sum_{n=0} \sum_{z \in E} |p^n(x, z) - p^n(y, z)| < \infty.$$

Moreover for all $x, y, z \in E_\pi$,

$$\sum_{n=0}^{\infty} (p^n(x, z) - p^n(y, z)) = (\mathbb{E}_y T_z - \mathbb{E}_x T_z) \pi(z).$$

(*d*) (The reflected random walk). Suppose that $\mathbb{E} z_1 < 0$. Then the reflected random walk (W_n) is ergodic (cf. Example 5.2(*d*)) and for any initial state $W_0 = w$, $\mathscr{L}(W_n) \to \pi$ in total variation norm as $n \to \infty$.

(*e*) (The renewal process on $\mathbb{R}_+$; cf. Example 6.1(*e*).) If F is spread-out and M_F is finite then for any constant $0 < c < \infty$;

$$\lim_{t\to\infty} \sup_{A \in \mathscr{R}_+ \cap [0,c]} |u(t + A) - M_F^{-1} \ell(A)| = 0; \tag{6.21}$$

moreover, for any constant $0 < \gamma < \infty$:

$$\int_{\mathbb{R}_+} |u(\mathrm{d}t) - u(\mathrm{d}t - \gamma)| < \infty.$$

6.4 Ergodicity of degree 2

An ergodic Markov chain which is recurrent of degree 2 is called *ergodic of degree* 2. Recall from Proposition 5.16 that recurrence of degree 2 means that the invariant probability distribution π is regular. Thus we obtain as an immediate consequence of the results of the preceding section the following:

Corollary 6.9. Suppose that (X_n) is ergodic of degree 2

(i) If λ is regular then

$$\sum_0^\infty \iint \lambda(\mathrm{d}x)\pi(\mathrm{d}y)\,\|P^n(x,\cdot)-P^n(y,\cdot)\| < \infty,$$

$$\sum_0^\infty \|\lambda P^n - \pi\| < \infty,$$

$$\lim_{n\to\infty} n\,\|\lambda P^n - \pi\| = 0$$

and

$$\sum_0^\infty (\lambda P^n f - \pi(f)) = \lambda(I - 1\otimes\pi)\bar{G}_{m_0,s,\nu}(I - 1\otimes\pi)f \quad \text{for all } f\in b\mathscr{E}.$$

(ii) For all $x\in R_1$:

$$\sum_0^\infty \|P^n(x,\cdot) - \pi\| < \infty$$

and

$$\lim_{n\to\infty} n\,\|P^n(x,\cdot)-\pi\| = 0. \quad \square$$

In addition to the above results we can prove that, if (X_n) is ergodic of degree 2, then the rate of convergence in (6.17) is n^{-2} whenever the initial distributions λ and μ are regular of degree 2:

Theorem 6.13. Suppose that the Markov chain (X_n) is ergodic of degree 2.

(i) If λ and μ are regular of degree 2 then

$$\sum_1^\infty n \iint \lambda(\mathrm{d}x)\mu(\mathrm{d}y)\,\|P^n(x,\cdot)-P^n(y,\cdot)\| < \infty, \tag{6.22}$$

$$\sum_1^\infty n\,\|\lambda P^n - \mu P^n\| < \infty$$

and

$$\lim_{n\to\infty} n^2\,\|\lambda P^n - \mu P^n\| = 0.$$

(ii) For all x, y belonging to the full set $R^{(2)}$,

$$\sum_1^\infty n \| P^n(x,\cdot) - P^n(y,\cdot) \| < \infty$$

and

$$\lim_{n\to\infty} n^2 \| P^n(x,\cdot) - P^n(y,\cdot) \| = 0.$$

Proof. We shall prove only (6.22), since the other results are immediate consequences of it. It is sufficient to consider only the case $m_0 = 1$ (cf. the proof of Theorem 6.11).

By the inequality (6.20)

$$\sum_1^\infty n \int \lambda(\mathrm{d}x) \| P^n(x,\cdot) - \nu P^{n-1} \|$$

$$\leq \sum_1^\infty n \mathbb{P}_\lambda\{T_\alpha \geq n\} + \sum_1^\infty n \int \lambda(\mathrm{d}x) |a(x) * u - u| * \sigma(E)_{n-1}$$

$$= \mathbb{E}_\lambda[\tfrac{1}{2} T_\alpha(T_\alpha + 1)] + \sum_1^\infty m \int \lambda(\mathrm{d}x) |a(x) * u_m - u_m| \sum_0^\infty \sigma_n(E)$$

$$+ \sum_1^\infty \lambda(\mathrm{d}x) |a(x) * u_m - u_m| \sum_1^\infty n\sigma_{n-1}(E).$$

By Proposition 5.17 the expectation $\mathbb{E}_\lambda T_\alpha^2$ is finite. Also $M_b^{(2)} = \mathbb{E}_\alpha S_\alpha^2$ is finite. Consequently,

$$\sum_1^\infty n\sigma_{n-1}(E) = \sum_1^\infty n \mathbb{P}_\alpha\{S_\alpha \geq n\} < \infty,$$

and by Theorem 6.4(ii),

$$\sum_1^\infty m \int \lambda(\mathrm{d}x) |a(x) * u_m - u_m|$$

$$\leq \tfrac{1}{2}(\mathbb{E}_\lambda T_\alpha^2 - \mathbb{E}_\lambda T_\alpha) \mathrm{Var}(u) + \mathbb{E}_\lambda T_\alpha \mathrm{Var}^{(2)}(u) < \infty.$$

Example 6.3. (*e*) (The renewal process). Suppose that F is spread-out and that $M_F^{(2)} = \mathbb{E}[z_1^2] = \int t^2 F(\mathrm{d}t)$ is finite. Then the convergence in (6.21) has rate t^{-1}.

We leave it to the reader to work out the other examples.

6.5 Geometric ergodicity

A geometrically recurrent, ergodic Markov chain is called *geometrically ergodic.*

We shall show that, if the Markov chain (X_n) is geometrically ergodic, then for π-almost all $x \in E$ the n-step transition probabilities $P^n(x, A)$ tend to their stationary limits $\pi(A)$ with a (common) geometric rate and uniformly over $A \in \mathscr{E}$. In fact, we have the following result:

Theorem 6.14. Suppose that (X_n) is ergodic. Each of the following three conditions is equivalent to the geometric ergodicity of (X_n):

(i) The embedded renewal sequence

$$u_0 = 1,\ u_n = \nu P^{(n-1)m_0} s \quad \text{for } n \geq 1,$$

is geometrically ergodic.

(ii) For some small function $s' \in \mathscr{S}^+$, some set $B \in \mathscr{E}^+$, there exist functions $M, \rho \in \mathscr{E}_+$, such that $M < \infty$ and $\rho < 1$ on B, and

$$|P^n s'(x) - \pi(s')| \leq M(x)(\rho(x))^n \quad \text{for all } x \in B,\ n \geq 0.$$

(iii) There is a function $M \in \mathscr{L}^1_+(\pi)$ and a constant $\rho < 1$ such that

$$\| P^n(x, \cdot) - \pi \| \leq M(x)\rho^n \quad \text{for all } x \in E,\ n \geq 0.$$

In the proof we need the following:

Lemma 6.2. Suppose that $m_0 = 1$, i.e., (s, ν) is an atom. Let $G_\alpha^{(r)} = G_{s,\nu}^{(r)} = \sum_0^\infty r^n (P - s \otimes \nu)^n$. Then, for any constant $r > 1$, function $f \in \mathscr{E}_+$, measure $\lambda \in \mathscr{M}_+$:

$$\lambda G_\alpha^{(r)} s < \infty \quad \text{implies } \lambda G_\alpha^{(r)} 1 < \infty,$$

and

$$\nu G_\alpha^{(r)} f < \infty \quad \text{implies } \pi G_\alpha^{(r)} f < \infty.$$

Proof. By the definition of the kernel $G_\alpha^{(r)}$ and by Proposition 4.8 we have

$$\begin{aligned}
\lambda G_\alpha^{(r)} 1 &= \sum_{m=0}^\infty \sum_{n=0}^\infty r^m \lambda (P - s \otimes \nu)^m (P - s \otimes \nu)^n s \\
&= \sum_{n=0}^\infty r^{-n} \sum_{m=n}^\infty r^m \lambda (P - s \otimes \nu)^m s \\
&\leq (r-1)^{-1} \lambda G_\alpha^{(r)} s,
\end{aligned}$$

which proves the first implication. The proof of the second is similar. □

Proof of Theorem 6.14. We start by proving that (i) implies (iii). Suppose for a while that $m_0 = 1$, and let $-1 \leq f \leq 1$ be arbitrary. By Theorem 4.1 and by (5.7)

$$\begin{aligned}
&|P^n f(x) - \pi(f)| \\
&\quad = |\mathbb{P}_x\{T_\alpha \geq n\} + a(x) * u * \sigma(f)_{n-1} - \pi(s) \sum_1^\infty \sigma_{m-1}(f)| \\
&\quad \leq \mathbb{P}_x\{T_\alpha \geq n\} + |a(x) * u - \pi(s) 1| * \sigma(E)_{n-1} + \pi(s) \sum_n^\infty \sigma_m(E).
\end{aligned}$$

Since $\sum_0^\infty a_m(x) = 1$ and $\pi(s) = M_b^{-1}$, we obtain further, writing $A_n(x) = \sum_{n+1}^\infty a_m(x)$.

$$|a(x)*u_n - \pi(s)1| \le a(x)*|u - M_b^{-1}|_n + \pi(s)A_n(x).$$

Consequently,

$$\int \pi(dx)\,\|P^n(x,\cdot) - \pi\| < \mathbb{P}_\pi\{T_\alpha \ge n\}$$
$$+ \pi(a)*|u - M_b^{-1}|*\sigma(E)_{n-1}$$
$$+ \pi(s)\pi(A)*\sigma(E)_{n-1} + \pi(s)\sum_n^\infty \sigma_m(E),$$

from which it follows that for all $r > 1$,

$$\sum_0^\infty r^n \int \pi(dx)\,\|P^n(x,\cdot) - \pi\| \le \pi G_\alpha^{(r)} 1$$
$$+ r\pi G_\alpha^{(r)} s \sum_{m=0}^\infty r^m |u_m - M_b^{-1}| v G_\alpha^{(r)} 1$$
$$+ r\pi G_\alpha^{(r)} s v G_\alpha^{(r)} 1 + \pi G_\alpha^{(r)} 1.$$

Now by Theorem 6.6 and by Lemma 6.2 there is a constant $r > 1$ such that the right hand side is finite. Set $M(x) = \sum_0^\infty r^n \|P^n(x,\cdot) - \pi\|$ and $\rho = r^{-1}$ to obtain (iii).

In the case of an arbitrary m_0 the implication (i)$\Rightarrow$(iii) follows by considering the m_0-step chain (X_{nm_0}) and by using the contractivity of P.

Trivially, (iii) implies both (i) and (ii).

Next we shall prove that (ii) implies (iii). Let $B' \in \mathscr{E}^+ \cap B$ be such that $M' = \sup_{x \in B'} M(x) < \infty$ and $\rho' = \sup_{x \in B'} \rho(x) < 1$. Let $m_0' \ge 1$, $\beta' > 0$ and $v' \in \mathscr{M}^+$ be such that the minorization condition $M(m_0', \beta', s', v')$ holds. Without any loss of generality we can assume that $\beta' = 1$ and that v' is a probability measure concentrated on B' (see Remarks 2.1(ii) and (iii)). Let

$$u_0' = 1, u_n' = v' P^{(n-1)m_0'} s' \quad \text{for } n \ge 1,$$

be the embedded renewal sequence associated with the minorization condition $M(m_0', 1, s', v')$. By the hypothesis

$$|u_n' - u_\infty'| \le \int_{B'} v'(dx)\,|P^{(n-1)m_0'} s'(x) - \pi(s')|$$
$$\le M'(\rho')^{(n-1)m_0'},$$

i.e., the embedded renewal sequence u' is geometrically ergodic. Since (i) always implies (iii), regardless of the particular choice of the embedded renewal sequence, the proof of the implication (ii)$\Rightarrow$(iii) is completed.

It remains to prove that geometric ergodicity is equivalent to (i). But this is a direct consequence of Proposition 5.20 and Lemma 5.7. □

Example 6.4. (*e*) (The renewal process). Suppose that F is spread-out and $\mathbb{E}e^{\gamma z_1} = \int e^{\gamma t}F(dt)$ is finite for some constant $\gamma > 0$. Then the convergence in (6.21) has rate $e^{-\gamma' t}$, for some constant $\gamma' > 0$.

6.6 Uniform ergodicity

An aperiodic, uniformly recurrent Markov chain is called *uniformly ergodic*.

We shall show that for a uniformly ergodic chain the operator norms

$$\| P^n - 1 \otimes \pi \| = \sup_{x \in E} \| P^n(x, \cdot) - \pi \|$$

converge to zero as $n \to \infty$. Note that, by contractivity, the convergence automatically has geometric rate. It also turns out that uniform ergodicity can be characterized as the smallness of the state space E.

Theorem 6.15. Either of the following two conditions is equivalent to the uniform ergodicity of (X_n):

(i) There exist a probability measure π on $(E, \mathscr{E})$, and constants $M < \infty$ and $\rho < 1$ such that

$$\| P^n - 1 \otimes \pi \| \le M\rho^n.$$

(ii) P is stochastic and aperiodic, and the state space E is small.

Proof. Suppose first that (X_n) is uniformly ergodic. There is no loss of generality in assuming that $m_0 = 1$, i.e. (s, v) is an atom. Since uniform recurrence means that the state space E is special, Proposition 5.13(iii) states that $\mathbb{E}_x T_\alpha = G_\alpha 1(x) - 1$ is bounded. Hence

$$\lim_{n \to \infty} \sup_{x \in E} \mathbb{P}_x\{T_\alpha \ge n\} = 0. \tag{6.23}$$

Consequently, by Theorem 6.5

$$\lim_{n \to \infty} \sup_{x \in E} | a(x) * u_n - \pi(s) | = 0. \tag{6.24}$$

Using Theorem 4.1 we obtain

$$\| P^n - 1 \otimes \pi \| \le \sup_{x \in E} \mathbb{P}_x\{T_\alpha \ge n\}$$
$$+ \sup_{x \in E} |a(x) * u - \pi(s) 1 | * \sigma(E)_{n-1} + \pi(s) \sum_n^{\infty} \sigma_m(E).$$

By (6.23) and (6.24) the right hand side tends to zero as $n \to \infty$. Thus we have (i).

If (i) holds, then (X_n) is ergodic by Proposition 6.3, and

$$\begin{aligned} P^{n+m_0} &\geq P^n s \otimes \nu \\ &\geq \tfrac{1}{2}\pi(s) 1 \otimes \nu \quad \text{for } n \text{ sufficiently big;} \end{aligned}$$

i.e., we have (ii).

That (ii) implies uniform ergodicity follows by using criterion (iii) of Proposition 5.23. □

Remarks 6.1. (i) That E is small, means the following: There exist an integer $m_0 \geq 1$, a constant $\beta > 0$ and a probability measure $\nu \in \mathcal{M}^+$ such that

$$P^{m_0}(x, \cdot) \geq \beta \nu(\cdot) \quad \text{for all } x \in E.$$

(ii) For a uniformly ergodic Markov chain the operator $\bar{G}_{m_0, \beta 1, \nu} = \sum_{n=0}^{\infty} (P^{m_0} - \beta 1 \otimes \nu)^n (I + \cdots + P^{m_0 - 1})$ is bounded; its norm $\| \bar{G}_{m_0, \beta 1, \nu} \|$ equals

$$\bar{G}_{m_0, \beta 1, \nu} 1 \equiv m_0 \beta^{-1}.$$

Note that the series $\sum_{n=0}^{\infty} (P^n - 1 \otimes \pi) = \sum_{n=0}^{\infty} (P - 1 \otimes \pi)^n$ converges absolutely (in the operator norm $\| \cdot \|$); by Corollary 6.9 its sum equals

$$\sum_{n=0}^{\infty} (P^n - 1 \otimes \pi) = (I - 1 \otimes \pi) \bar{G}_{m_0, \beta 1, \nu} (I - 1 \otimes \pi).$$

6.7 *R*-ergodic kernels

We shall now consider a general R-recurrent kernel K. The kernel K is called *R-ergodic*, if it is aperiodic and R-positive recurrent. By Proposition 5.4 the transformed kernel

$$\tilde{K} = R I_{h^{-1}} K I_h,$$

where h is the (essentially unique) minimal R-invariant function, is the transition probability of an ergodic Markov chain $(\tilde{X}_n)$ with state space $\tilde{E} = \{h < \infty\}$.

If K is R-ergodic, then by Corollary 6.3 and Proposition 6.2

$$\lim_{n \to \infty} \| R^n K^n(x, \cdot) - h(x)\pi \|_h = 0 \quad \text{for all } x \in \tilde{E}$$

and

$$\lim_{n \to \infty} R^n K^n f(x) = \infty \quad \text{for all } x \in (\tilde{E})^c,\ f \in \mathcal{E}^+.$$

Hence, in particular,

$$\liminf_{n \to \infty} R^n K^n f(x) > 0 \quad \text{for all } x \in E,\ f \in \mathcal{E}^+.$$

There is also a converse result to this:

Proposition 6.4. Suppose that for some constant $0 < R < \infty$, some σ-finite measure $\varphi \in \mathcal{M}^+$:

(i) For all $0 \le r < R$,

$$\lim_{n\to\infty} r^n K^n(x, A) = 0 \quad \text{for some } x \in E, \text{ some } \varphi\text{-positive } A \in \mathscr{E};$$

and

(ii) $$\liminf_{n\to\infty} R^n K^n(x, A) > 0 \quad \text{for all } x \in E, \text{ all } \varphi\text{-positive } A \in \mathscr{E}.$$

Then the kernel K is R-ergodic.

Proof. By (ii) K is φ-irreducible and aperiodic. (i) and (ii) together imply that R is the convergence parameter of K, and that K is R-recurrent. By Theorem 6.9 K is R-positive recurrent. □

The 'transformation' of Corollary 6.8(ii) from the Markov chain $(\tilde{X}_n)$ to the kernel K leads to the following:

Corollary 6.10. Suppose that K is R-ergodic. Then for π-almost all $x, y \in E$:

$$\sum_{n=0}^{\infty} R^n \| h(x)^{-1} K^n(x, \cdot) - h(y)^{-1} K^n(y, \cdot) \|_h < \infty. \quad \square$$

We call K geometrically (resp. uniformly) R-ergodic, if it is aperiodic and geometrically (resp. uniformly) recurrent. It follows from Proposition 5.25 and Corollary 5.7 that an irreducible kernel K is geometrically (resp. uniformly) R-ergodic if and only if the Markov chain $(\tilde{X}_n)$ is geometrically (resp. uniformly) ergodic.

The 'transformation' of Theorems 6.14 and 6.15 gives us the following two corollaries:

Corollary 6.11. Suppose that K is irreducible. Either of the following two conditions is equivalent to the geometric R-ergodicity of K:

(i) K is aperiodic and the renewal sequence

$$\tilde{u}_0 = 1, \tilde{u}_n = R^{nm_0} \nu K^{(n-1)m_0} s \quad \text{for } n \ge 1,$$

is geometrically ergodic.

(ii) K is R-ergodic, and there exist a function $M \in \mathscr{L}^1_+(\pi)$ and a constant $\rho < 1$ such that

$$\| R^n K^n(x, \cdot) - h(x)\pi \|_h \le M(x)\rho^n \quad \text{for all } x \in E, n \ge 0. \quad \square$$

If K is uniformly ergodic then even the operator norms

$$\| R^n K^n - h \otimes \pi \|_h \overset{\text{def}}{=} \sup_{x \in E} \| R^n h(x)^{-1} K^n(x, \cdot) - \pi \|_h$$

converge:

Corollary 6.12. Either of the following two conditions is equivalent to the uniform R-ergodicity of K:

(i) K is R-ergodic and there exist constants $M < \infty$ and $\rho < 1$ such that

$$\| R^n K^n - h \otimes \pi \|_h \leq M \rho^n \quad \text{for all } n \geq 0.$$

(ii) K is R-ergodic and the R-invariant function h is small, i.e., there exist an integer $m_0 \geq 1$, constant $\beta > 0$ and measure $\nu \in \mathcal{M}^+$ such that

$$K^{m_0} \geq \beta h \otimes \nu. \qquad \square$$

As an application of the concept of R-ergodicity we shall consider the so-called quasi-stationary distributions of an irreducible Markov chain (X_n) with transition probability P. The following considerations are of interest only if P is properly substochastic, i.e., $\{P1 < 1\} \in \mathscr{E}^+$.

Fix a set $B \in \mathscr{E}^+$. For any $n \geq 0$, $x \in E$, $A \in \mathscr{E} \cap B$, the ratio $P^n(x, A)/P^n(x, B)$ can be interpreted as the conditional probability

$$\frac{P^n(x, A)}{P^n(x, B)} = \mathbb{P}_x\{X_n \in A \mid X_n \in B\}. \tag{6.25}$$

If there is a full set $F \in \mathscr{E}$ such that these conditional probabilities tend to a limit as $n \to \infty$, which is independent of the initial state $x \in F$, and if the limit, considered as a function of A, is a probability measure on $(B, \mathscr{E} \cap B)$, then the limit is called the *quasi-stationary distribution on B*.

Using Corollary 6.3 we obtain a set of sufficient conditions for the existence of the quasi-stationary distribution on the given set B:

Corollary 6.13. Suppose that the transition probability P is R-ergodic for some $R \geq 1$. Let h be the (essentially unique) minimal R-invariant function and let π be the (essentially unique) R-invariant measure for P. If

$$0 < \pi(B) < \infty \quad \text{and} \quad \inf_B h > 0,$$

then the quasi-stationary distribution on B exists and equals $\pi(A)/\pi(B)$, $A \in \mathscr{E} \cap B$. Moreover the convergence of (6.25) to its limit $\pi(A)/\pi(B)$ is uniform over $A \in \mathscr{E} \cap B$, for any $x \in F = \tilde{E}$. $\square$

Note that if $B \in \mathscr{E}^+$ is a small set then it satisfies the hypotheses of the above corollary. Hence in particular, if P is R-ergodic and the whole state space E is small, then the quasi-stationary distribution on E exists: for all $x \in \tilde{E}$,

$$\lim_{n \to \infty} \mathbb{P}_x\{X_n \in A \mid L_E \geq n\} = \pi(A)/\pi(E) \quad \text{uniformly over } A \in \mathscr{E}.$$

(Here $L_E \overset{\text{def}}{=} \sup\{n \geq 0 : X_n \in E\}$ is the lifetime of the Markov chain (X_n).)

7

Miscellaneous limit theorems for Harris recurrent Markov chains

Throughout this last chapter we assume (without further mentioning) *that $K = P$ is the transition probability of an aperiodic, Harris recurrent Markov chain $(X_n; n \geq 0)$.*

We assume that P satisfies the minorization condition $M(m_0, 1, s, \nu)$. We fix an initial distribution λ and a π-integrable function f.

In Sections 7.1 and 7.2 we shall examine the sums of transition probabilities $\sum_{n=0}^{\infty} \lambda P^n f$. In Section 6.3 we considered the positive recurrent case only; now we shall not exclude null recurrence. Section 7.1 deals with the differences $\sum_{n=0}^{\infty} (\lambda P^n f - \nu P^n f)$ whereas Section 7.2 is concerned with the convergence of the ratios $\sum_0^N \lambda P^n f / \sum_0^N \nu P^n s$, $N \to \infty$.

In Section 7.3 we shall study the convergence of the individual ratios $\lambda P^n f / \nu P^n s$.

Finally, in Section 7.4 we shall prove a central limit theorem for the sums $\sum_0^N f(X_n)$.

7.1 Sums of transition probabilities

In the following theorem sufficient conditions are given for the convergence of the series

$$\sum_{n=0}^{\infty} (\lambda P^n f - \nu P^n f).$$

Recall the definition of the kernel $\bar{G}_{m_0,s,\nu} = \sum_{n=0}^{\infty} (P^{m_0} - s \otimes \nu)^n (I + \cdots + P^{m_0 - 1})$. Also recall from Proposition 5.13 that λ is $|f|$-regular, if and only if $\lambda \bar{G}_{m_0,s,\nu} |f|$ is finite.

Theorem 7.1. If $\pi(f) = 0$, λ is $|f|$-regular, and

$$\sup_{N \geq 0} \left| \sum_0^N \nu P^n f \right| < \infty, \tag{7.1}$$

then

$$\sum_0^{\infty} (\lambda P^n f - \nu P^n f) = \lambda \bar{G}_{m_0,s,\nu} f. \tag{7.2}$$

In particular, if $\pi(f) = 0$ and (7.1) holds, then there is a full set $F (= R_{|f|})$

such that

$$\sum_0^\infty (P^n f(x) - \nu P^n f) = \bar{G}_{m_0,s,\nu} f(x) \quad \text{for all } x \in F.$$

The proof is based on the following:

Lemma 7.1. Let $a = (a_n; n \geq 0)$ be a probability distribution on $\mathbb{N}$ and let $c = (c_n; n \geq 0)$ be an arbitrary bounded sequence satisfying

$$\lim_{n \to \infty} (c_n - c_{n-1}) = 0.$$

Then

$$\lim_{n \to \infty} (a * c_n - c_n) = 0.$$

Proof. For any $0 \leq N \leq n$, write

$$a * c_n - c_n = \sum_{m=0}^{N} a_m (c_{n-m} - c_n)$$

$$+ \sum_{m=N+1}^{n} a_m c_{n-m} - \sum_{m=N+1}^{\infty} a_m c_n,$$

and let first n and then N tend to ∞. □

Proof of Theorem 7.1. Suppose that $m_0 = 1$. Note that $\lim_{N \to \infty} \nu P^N f = 0$. Consider the difference

$$\sum_0^N \lambda P^n f - \sum_0^{N-1} \nu P f = \mathbb{E}_\lambda \sum_0^{T_\alpha \wedge N} f(X_n)$$

$$+ \lambda(a) * u * \sigma(f) * 1_{N-1} - u * \sigma(f) * 1_{N-1}.$$

As $N \to \infty$ the first term on the right hand side tends to the finite limit $\lambda G_\alpha f = \lambda G_{s,\nu} f$. In order to see that the second term tends to zero set $a = \lambda(a)$ and $c = u * \sigma(f) * 1$ in Lemma 7.1.

The proof of the general case, m_0 arbitrary, follows by considering the m_0-step chain (X_{nm_0}) and the function $f + \cdots + P^{m_0 - 1} f$. □

Corollary 7.1. Suppose that $\pi(f) = 0$, λ is $|f|$-regular and the series $\sum_0^\infty \nu P^n f$ converges. Then the series $\sum_0^\infty \lambda P^n f$ converges and

$$\sum_0^\infty \lambda P^n f = \sum_0^\infty \nu P^n f + \lambda \bar{G}_{m_0,s,\nu} f.$$

In particular, if $\pi(f) = 0$ and $\sum_0^\infty \nu P^n f$ converges, then

$$\sum_0^\infty P^n f(x) = \sum_0^\infty \nu P^n f + \bar{G}_{m_0,s,\nu} f(x) \quad \text{for all } x \in R_{|f|}.$$ □

It turns out that, if the function $f \in \mathscr{L}_0^1(\pi)$ is special, then the hypothesis (7.1) of Theorem 7.1 holds true.

We denote by $\|\cdot\|$ the *supremum norm* in the space $b\mathscr{E}$ of bounded measurable functions on $(E, \mathscr{E})$:

$$\|f\| = \sup_E |f|, \quad f \in b\mathscr{E}.$$

Theorem 7.2. If f is special and $\pi(f) = 0$, then

$$\sup_{N \geq 0} \left\| \sum_0^N P^n f \right\| < \infty, \tag{7.3}$$

and (7.2) holds for any initial distribution λ.

Proof. Without any loss of generality we may assume that $m_0 = 1$ (cf. the proof of Theorem 7.1).

By hypothesis

$$\sum_0^\infty \sigma_n(f) = \pi(s)^{-1} \pi(f) = 0.$$

Set $B_n(f) = \sum_{m=n}^\infty \sigma_m(f) = v(P - s \otimes v)^n G_\alpha f$. It follows that

$$\sigma(f) * 1_{n-1} = -B_n(f) = -v(P - s \otimes v)^n G_\alpha f \quad \text{for all } n \geq 0,$$

and hence

$$\begin{aligned} |B_n(f)| &\leq \|G_\alpha f\| v(P - h \otimes v)^n 1 \\ &= \|G_\alpha f\| \mathbb{P}_\alpha\{S_\alpha \geq n+1\} \\ &= \|G_\alpha f\| B_n \quad \text{for all } n \geq 0 \text{ (see (4.22))}, \end{aligned}$$

and

$$\begin{aligned} \left| \sum_0^{N-1} vP^n f \right| \leq |u * \sigma(f) * 1_{N-1}| &= |u * B(f)_N| \\ &\leq \|G_\alpha f\| u * B_N \\ &= \|G_\alpha f\| \quad \text{(recall (4.6))}. \end{aligned}$$

The assertion (7.3) follows from the estimates

$$\begin{aligned} \left| \sum_0^N P^n f(x) \right| &= \left| \mathbb{E}_x \sum_0^{T_\alpha \wedge N} f(X_n) + a(x) * u * \sigma(f) * 1_{N-1} \right| \\ &\leq G_\alpha |f|(x) + \|G_\alpha f\|. \end{aligned}$$

The second assertion (7.2) is a direct consequence of Theorem 7.1. □

Example 7.1. (*a*) Suppose that P is an aperiodic Harris recurrent transition matrix (see Examples 3.4(*a*) and 5.1(*a*)). Then for all $x, y, z \in E$, all $x_0 \in E_\pi$:

$$\sup_{N \geq 0} \sup_{x \in E} \left| \sum_0^N (\pi(z) P^n(x, y) - \pi(y) P^n(x, z)) \right| < \infty$$

and (writing $G_{x_0}(x,y) = \mathbb{E}_x \sum_0^{T_{x_0}} 1_{\{X_n = y\}}$)

$$\sum_{n=0}^{\infty} [\pi(z)(P^n(x,y) - P^n(x_0,y)) - \pi(y)(P^n(x,z) - P^n(x_0,z))]$$
$$= \pi(z)G_{x_0}(x,y) - \pi(y)G_{x_0}(x,z).$$

7.2 Ratios of sums of transition probabilities

We shall start by proving a ratio limit theorem for renewal sequences.

Let $u = (u_n; n \geq 0)$ be an aperiodic, recurrent renewal sequence and let $a = (a_n; n \geq 0)$ be an arbitrary delay distribution. If u is positive recurrent, then by Theorem 6.2, the expectation of the mean number of renewal epochs during the interval $[0, N]$, that is

$$\mathbb{E}_a\left[(N+1)^{-1}\sum_0^N 1_{\{Y_n = 1\}}\right] = (N+1)^{-1}\sum_0^N a * u_n,$$

tends to the limit $u_\infty = M_b^{-1} > 0$ as $N \to \infty$. It follows that the ratios

$$\frac{\mathbb{E}_a \sum_0^N 1_{\{Y_n = 1\}}}{\mathbb{E}_\delta \sum_0^N 1_{\{Y_n = 1\}}} = \frac{\sum_0^N a * u_n}{\sum_0^N u_n} \tag{7.4}$$

tend to the limit 1 as $N \to \infty$.

The ratios (7.4) converge also in the case of null recurrence:

Proposition 7.1. Suppose that u is an aperiodic, recurrent renewal sequence. Then for any delay distribution a,

$$\lim_{N \to \infty} \frac{\sum_0^N a * u_n}{\sum_0^N u_n} = 1.$$

Proof. The ratio (7.4) can be written in the form

$$\frac{a * u * 1_N}{u * 1_N} = \frac{u * 1_N - A * u_N}{u * 1_N}.$$

(Here $A_n = \sum_{n+1}^{\infty} a_m$ as before.) Hence, what we have to prove is

$$\lim_{N \to \infty} \frac{A * u_N}{u * 1_N} = 0.$$

Let $N_0 \le N$ be a fixed integer. We have

$$\frac{A*u_N}{u*1_N} \le \frac{\sum_{n=0}^{N_0-1} A_n u_{N-n}}{\sum_{n=0}^{N} u_n} + \frac{A_{N_0} \sum_{n=N_0}^{N} u_{N-n}}{\sum_0^N u_n}.$$

Now let N and N_0 tend to ∞, and use the fact that $\sum_0^\infty u_n = \infty$ by recurrence. □

For a positive Harris recurrent Markov chain (X_n) the ratios

$$\frac{\sum_0^N \lambda P^n f}{\sum_0^N \nu P^n s}$$

tend to $\pi(f)/\pi(s)$ as $N \to \infty$. The following theorem states that under some additional hypotheses this result holds true also in the null recurrent case:

Theorem 7.3. If λ is $|f|$-regular, then

$$\lim_{N\to\infty} \frac{\sum_0^N \lambda P^n f}{\sum_0^N \nu P^n s} = \frac{\pi(f)}{\pi(s)}. \tag{7.5}$$

Proof. Without any loss of generality we may suppose that $m_0 = 1$. It also suffices to consider the case where the function f is non-negative.

By Theorem 4.1

$$\frac{\sum_0^N \lambda P^n f}{\sum_0^N u_n} = \frac{\mathbb{E}_\lambda \sum_0^{T_\alpha \wedge N} f(X_n)}{\sum_0^N u_n} + \frac{\sum_0^{N-1} \lambda(a)*u*\sigma(f)_n}{\sum_0^N u_n}.$$

The first term on the right hand side is dominated by $\lambda G_\alpha f/\sum_0^N u_n$ which by recurrence tends to zero as $N \to \infty$. By Proposition 7.1 the second term tends to the limit

$$\sum_0^\infty \lambda(a_m) \sum_0^\infty \sigma_n(f) = \pi(s)^{-1}\pi(f).$$

Consequently,

$$\lim_{N\to\infty} \frac{\sum_0^N \lambda P^n f}{\sum_0^N u_n} = \frac{\pi(f)}{\pi(s)},$$

from which the result follows. □

Corollary 7.2. (i) If f is special, then (7.5) holds for any initial distribution λ.

(ii) We have

$$\lim_{N\to\infty}\frac{\sum_0^N P^n f(x)}{\sum_0^N \nu P^n s}=\frac{\pi(f)}{\pi(s)}\quad\text{for } \pi\text{-almost all } x\in E. \qquad \square$$

Example 7.2. (*a*) Suppose that P is an aperiodic Harris recurrent transition matrix. Then for all $x, y\in E$, $z\in E_\pi$:

$$\lim_{N\to\infty}\frac{\sum_0^N P^n(x,y)}{\sum_0^N P^n(z,z)}=\frac{\pi(y)}{\pi(z)}.$$

7.3 Ratios of transition probabilities

In the following theorem sufficient conditions are given for the convergence of the ratios $\lambda P^n f/\nu P^n s$.

Theorem 7.4. Suppose that $f\in\mathscr{L}_+^1(\pi)$ and that the embedded renewal sequence $u_0=0$, $u_n=\nu P^{(n-1)m_0}s$ for $n\geq 1$, satisfies

$$\lim_{n\to\infty}\frac{u_{n+1}}{u_n}=1. \tag{7.6}$$

Then:

(i) We have

$$\liminf_{n\to\infty}\frac{\lambda P^{nm_0}f}{\nu P^{nm_0}s}\geq\frac{\pi(f)}{\pi(s)}.$$

(ii) If f is small and

$$\limsup_{n\to\infty}\frac{\lambda P^{nm_0}s}{\nu P^{nm_0}s}\leq 1, \tag{7.7}$$

then

$$\lim_{n\to\infty}\frac{\lambda P^n f}{\nu P^n s}=\frac{\pi(f)}{\pi(s)}. \tag{7.8}$$

(iii) If λ is small and

$$\limsup_{n\to\infty}\frac{\nu P^{nm_0}f}{\nu P^{nm_0}s}\leq\frac{\pi(f)}{\pi(s)},$$

then (7.8) holds.

(iv) If f and λ are small, then (7.8) holds.

Proof. (i) It suffices to consider the case $m_0=1$. For any sequence $c=$

$(c_n; n \geq 0)$, for any fixed $N \geq 0$, write $c^{(N)}$ for the truncated sequence

$$c_n^{(N)} = c_n 1_{\{n \leq N\}}.$$

By Theorem 4.1

$$P^n f = \lambda(a^{(N)}) * u * \sigma^{(N)}(f)_{n-1} + r_{n,N}(\lambda, f) \tag{7.9}$$

where $r_{n,N}(\lambda, f) \geq 0$. Hence, using (7.6), we have for any $N \geq 0$,

$$\begin{aligned}\liminf_{n\to\infty} \frac{\lambda P^n f}{\nu P^n s} &= \liminf_{n\to\infty} \frac{\lambda P^n f}{u_{n+1}} \\ &\geq \lim_{n\to\infty} \frac{\lambda(a^{(N)}) * u * \sigma^{(N)}(f)_{n-1}}{u_{n+1}} \\ &= \sum_{m=0}^{N} \lambda(a_m) \sum_{n=0}^{N} \sigma_n(f) \\ &\to \frac{\pi(f)}{\pi(s)} \quad \text{as} \quad N \to \infty.\end{aligned}$$

(ii) Assume first that $m_0 = 1$. We obtain for all $m \geq 0$:

$$\lim_{N\to\infty}\left(\lim_{n\to\infty} \frac{\lambda(a^{(N)}) * u * \sigma^{(N)}(P^m s)_{n-1}}{u_{n+1}}\right) = \frac{\pi(P^m s)}{\pi(s)} = 1.$$

Since by the hypotheses

$$\limsup_{n\to\infty} \frac{\lambda P^{n+m} s}{u_{n+1}} \leq 1 \quad \text{for any } m,$$

we have, by using (7.7) and (7.9) with $f = P^m s$,

$$\lim_{N\to\infty}\left(\lim_{n\to\infty} \frac{r_{n,N}(\lambda, P^m s)}{u_{n+1}}\right) = 0.$$

Since f is small, there exist an integer $m \geq 0$ and a constant $\gamma > 0$ such that

$$f \leq \gamma P^m s.$$

Consequently,

$$\lim_{N\to\infty}\left(\limsup_{n\to\infty} \frac{r_{n,N}(\lambda, f)}{u_{n+1}}\right) = 0,$$

and therefore

$$\lim_{n\to\infty} \frac{\lambda P^n f}{u_{n+1}} = \lim_{N\to\infty}\left(\lim_{n\to\infty} \frac{\lambda(a^{(N)}) * u * \sigma^{(N)}(f)_{n-1}}{u_{n+1}}\right) = \frac{\pi(f)}{\pi(s)}.$$

The assertion in the case $m_0 \geq 2$ follows after observing that for every $m \geq 0$, the function $P^m f$ is small for the m_0-step chain.

(iii) The proof is similar to that of part (ii).

(iv) The inequality (7.7) is trivially satisfied with $\lambda = \nu$. Hence by part (ii),

$$\lim_{n\to\infty} \frac{\nu P^n f}{\nu P^n s} = \frac{\pi(f)}{\pi(s)}.$$

The assertion now follows from part (iii). □

In the following theorem a necessary and sufficient condition is given for the convergence to be valid for all initial distributions λ and all small functions f.

Theorem 7.5. We have (7.8) for all λ and all small f if and only if

$$\limsup_{n\to\infty} \left\| \frac{P^n s}{\nu P^n s} \right\| < \infty. \tag{7.10}$$

Proof. Suppose first that (7.10) holds, and let $N, \gamma < \infty$ be such that

$$\gamma = \sup_{n \geq N} \left\| \frac{P^{nm_0} s}{\nu P^{nm_0} s} \right\| < \infty.$$

For any fixed integer $m \geq 0$, any probability measure λ,

$$\limsup_{n\to\infty} \left| \frac{\lambda P^{nm_0} s - \nu P^{nm_0} s}{\nu P^{nm_0} s} \right|$$

$$= \limsup_{n\to\infty} \left| (\lambda P^{mm_0} - \nu P^{mm_0}) \frac{P^{(n-m)m_0} s}{\nu P^{(n-m)m_0} s} \frac{u_{n-m+1}}{u_{n+1}} \right|$$

$$\leq \gamma \| \lambda P^{mm_0} - \nu P^{mm_0} \|.$$

Letting $m \to \infty$ and using Corollary 6.7(i) we immediately get (7.7). This implies (7.8).

Suppose now that (7.10) does not hold. Then for any $m \geq 1$, there exist $x_m \in E$, $n_m \geq 0$, such that

$$P^{n_m} s(x_m) \geq 2^{m+1} \nu P^{n_m} s.$$

Define a discrete probability measure λ on $(E, \mathscr{E})$ by setting $\lambda(\{x_m\}) = 2^{-m}$, $m \geq 1$. Then clearly,

$$\frac{\lambda P^{n_m} s}{\nu P^{n_m} s} \geq 2 \quad \text{for all } m \geq 1,$$

contradicting (7.8).

Example 7.3. (*a*) Suppose that P is an aperiodic, recurrent transition matrix. If

$$\lim_{n\to\infty} \frac{P^{n+1}(z,z)}{P^n(z,z)} = 1 \quad \text{for some } z \in E_\pi,$$

then

$$\lim_{n\to\infty}\frac{P^n(x,y)}{P^n(z,z)}=\frac{\pi(y)}{\pi(z)}$$

for all $x,y\in E_\pi$.

7.4 A central limit theorem

In this section we shall prove a central limit theorem for the functionals

$$\xi_N(f)\overset{\text{def}}{=}\sum_{n=0}^{N}f(X_n),\quad f\in\mathscr{L}^1(\pi).$$

Note that, when $f=1_A$, $\xi_N(A)\overset{\text{def}}{=}\xi_N(1_A)$ counts the number of visits by (X_n) to the set A during $[0,N]$. We write $\mathscr{N}(M,\sigma^2)$ for a normal random variable with mean M and variance σ^2.

Theorem 7.6. Suppose that (X_n) is ergodic. Then for any initial distribution λ, any $f\in\mathscr{L}_0^1(\pi)$ such that the measure $\pi I_{|f|}$ is $|f|$-regular,

$$N^{-1/2}\xi_N(f)\to\mathscr{N}(0,\sigma_f^2)\quad\text{in } \mathbb{P}_\lambda\text{-distribution as } N\to\infty,$$

where the variance σ_f^2 is given by

$$\sigma_f^2=\pi(f^2)+2m_0^{-1}\left[\sum_{m=1}^{m_0}(m_0-m)\pi I_fP^mf+\sum_{m=1}^{m_0}\pi I_fP^m\bar{G}_{m_0,s,v}f\right].$$

Proof. We fix λ and write simply $\mathbb{P}_\lambda=\mathbb{P}$.

Since the m_0-step chain $(X_{nm_0};n\geq 0)$ has the atom (s,v) we can construct its split chain $(X_{nm_0},Y_n;n\geq 0)$ in the manner described in Section 4.4. We shall next put the *entire* Markov chain (X_n) and the incidence sequence (Y_n) on the same probability space: This can be achieved by giving the conditional probabilities

$$\begin{aligned}\mathbb{P}\{Y_n=1,X_{nm_0+1}\in dx_1,\ldots,X_{nm_0+m_0-1}\in dx_{m_0-1},X_{(n+1)m_0}\in dy\\ |\mathscr{F}^X_{nm_0}\vee\mathscr{F}^Y_{n-1};X_{nm_0}=x\}\\ =\mathbb{P}_x\{Y_0=1,X_1\in dx_1,\ldots,X_{m_0-1}\in dx_{m_0-1},X_{m_0}\in dy\}\\ =r(x,y)P(x,dx_1)\ldots P(x_{m_0-1},dy),\quad n\geq 0,\quad x,y,\ldots\in E,\end{aligned}$$

where $r\in(\mathscr{E}\otimes\mathscr{E})_+$ is the Radon–Nikodym derivative

$$r(x,y)=\frac{s(x)v(dy)}{P^{m_0}(x,dy)}.$$

It is easy to see that, for every $q\geq 1$, the conditional distribution of the post-qm_0-chain $(X_{qm_0+n};n\geq 0)$ given $\mathscr{F}^X_{(q-1)m_0}\vee\mathscr{F}^Y_{q-1}$ and given $Y_{q-1}=1$ is the same as the $\mathbb{P}_v$-distribution of $(X_n;n\geq 0)$:

$$\begin{aligned}\mathscr{L}(X_{qm_0+n};n\geq 0|\mathscr{F}^X_{(q-1)m_0}\vee\mathscr{F}^Y_{q-1};Y_{q-1}=1)\\ =\mathscr{L}_v(X_n;n\geq 0).\end{aligned}\tag{7.11}$$

For every $N, i \geq 1$, let us write

$$i(N) = \sup\{i \geq 0 : (T_\alpha(i) + 1)m_0 \leq N\},$$

$$L(N) = T_\alpha(i(N)) = \sup\{n \geq 0 : (n+1)m_0 \leq N,\, Y_n = 1\},$$

$$\zeta(i) = \sum_{n=(T_\alpha(i-1)+1)m_0}^{(T_\alpha(i)+1)m_0-1} f(X_n) = \sum_{n=T_\alpha(i-1)+1}^{T_\alpha(i)} Z_n,$$

where

$$Z_n = \sum_{m=0}^{m_0-1} f(X_{nm_0+m}).$$

We can decompose the sum $\xi_N(f) = \sum_{n=0}^{N} f(X_n)$ as follows

$$\xi_N(f) = \sum_{n=0}^{(T_\alpha m_0 + m_0 - 1) \wedge N} f(X_n) + \sum_{i=1}^{i(N)} \zeta(i) + \sum_{n=(L(N)+1)m_0}^{N} f(X_n). \tag{7.12}$$

By (7.11) the classes $\{\zeta(j); 1 \leq j \leq i-2\}$ and $\{\zeta(j); j \geq i\}$ are independent for every $i \geq 3$. The (unconditional) distribution of every $\zeta(i)$, $i \geq 1$, is the same as the $\mathbb{P}_\alpha$-distribution of the functional $\sum_{n=m_0}^{S_\alpha m_0 + m_0 - 1} f(X_n)$: for every $i \geq 1$,

$$\mathscr{L}(\zeta(i)) = \mathscr{L}_\alpha\left(\sum_{n=m_0}^{S_\alpha m_0 + m_0 - 1} f(X_n)\right)$$

$$= \mathscr{L}_\nu\left(\sum_{n=0}^{T_\alpha m_0 + m_0 - 1} f(X_n)\right) \quad \text{by (7.11)},$$

$$= \mathscr{L}_\nu\left(\sum_{n=0}^{T_\alpha} Z_n\right). \tag{7.13}$$

Thus, for example,

$$\mathbb{E}\zeta(i) = \mathbb{E}_\nu\left[\sum_{n=0}^{T_\alpha}\sum_{m=0}^{m_0-1} f(X_{nm_0+m})\right]$$

$$= \sum_{n=0}^{\infty} \int \mathbb{E}_\nu\left[\sum_{m=0}^{m_0-1} f(X_{nm_0+m}); Y_0, \ldots, Y_{n-1} = 0; X_{nm_0} \in \mathrm{d}y\right]$$

$$= \pi(s)^{-1} \int \pi(\mathrm{d}y)\mathbb{E}_y \sum_{m=0}^{m_0-1} f(X_m)$$

by conditioning w.r.t. $\mathscr{F}^X_{nm_0} \vee \mathscr{F}^Y_{n-1}$ and using (7.11),

$$= (\pi(z))^{-1} m_0 \pi(f) = 0 \tag{7.14}$$

by the hypothesis.

The first term on the right hand side of (7.12) is dominated by

$$\sum_{n=0}^{T_\alpha m_0 + m_0 - 1} f(X_n)$$

which does not depend on N and therefore it is finite almost surely. Therefore, when divided by $N^{1/2}$, it tends to zero in probability as $N \to \infty$.

Let $c \geq 0$ be an arbitrary constant. For the third term we obtain (the notation $[t]$ means the integer part of the real number $t \in \mathbb{R}_+$):

$$\begin{aligned}
&\mathbb{P}\left\{\left|\sum_{n=(L(N)+1)m_0}^{N} f(X_n)\right| > c\right\} \\
&\leq \mathbb{P}\left\{\sum_{n=(L(N)+1)m_0}^{N} |f(X_n)| > c\right\} \\
&= \sum_{m=1}^{[N/m_0]} \mathbb{P}\left\{\sum_{n=([N/m_0]-m+1)m_0}^{N} |f(X_n)| > c; L(N) = [N/m_0] - m\right\} \\
&= \sum_{m=1}^{[N/m_0]} \mathbb{P}\{Y_{[N/m_0]-m} = 1\}\, \mathbb{P}_\alpha\left\{\sum_{n=m_0}^{N-([N/m_0]-m)m_0} |f(X_n)| > c; S_\alpha \geq m\right\} \\
&\leq \sum_{m=1}^{\infty} \mathbb{P}_\alpha\left\{\sum_{n=m_0}^{mm_0+m_0-1} |f(X_n)| > c; S_\alpha \geq m\right\} \\
&\leq \mathbb{E}_\alpha S_\alpha < \infty \text{ by positive recurrence.}
\end{aligned}$$

Hence, by the monotone convergence theorem

$$\lim_{c \uparrow \infty} \downarrow \sum_{m=1}^{\infty} \mathbb{P}_\alpha\left\{\sum_{n=m_0}^{mm_0+m_0-1} |f(X_n)| > c; S_\alpha \geq m\right\} = 0.$$

It follows that

$$N^{-1/2} \sum_{n=(L(N)+1)m_0}^{N} f(X_n) \to 0 \text{ in probability as } N \to \infty.$$

Consequently, it is sufficient to prove that

$$N^{-1/2} \sum_{i=1}^{i(N)} \zeta(i) \to \mathcal{N}(0, \sigma_f^2) \text{ in distribution as } N \to \infty. \tag{7.15}$$

We shall first prove the asymptotic normality of the random variables

$$I^{-1/2} \sum_{i=1}^{I} \zeta(i) \quad \text{as} \quad I \to \infty.$$

Let $m \geq 2$ be an arbitrary fixed integer. We can write

$$\sum_{i=1}^{I} \zeta(i) = \sum_{j=0}^{[I/m]-1} \eta(j) + \sum_{j=1}^{[I/m]-1} \zeta(jm) + \sum_{i=[I/m]m}^{I} \zeta(i) \tag{7.16}$$

where the random variables

$$\eta(j) \overset{\text{def}}{=} \zeta(jm+1) + \zeta(jm+2) + \cdots + \zeta(jm+m-1), \quad j \geq 0,$$

are i.i.d. with common mean

$$\mathbb{E}\eta(j) = \mathbb{E}\eta(0) = (m-1)\mathbb{E}\zeta(1) = 0 \qquad \text{(see (7.14))}$$

and variance

$$\sigma_m^2 = \mathbb{E}\eta(0)^2 = \mathbb{E}(\zeta(1) + \cdots + \zeta(m-1))^2$$
$$= (m-1)\mathbb{E}\zeta(1)^2 + 2(m-2)\mathbb{E}\zeta(1)\zeta(2),$$

and where the random variables $\zeta(jm)$, $j \geq 1$, are i.i.d. with zero mean and variance

$$\mathbb{E}\zeta(jm)^2 = \mathbb{E}\zeta(1)^2.$$

Now divide both sides of (7.16) by $I^{1/2}$ and let $I \to \infty$ (m is fixed). By the central limit theorem for i.i.d. random variables the first term on the right hand side tends to

$$\mathcal{N}(0, m^{-1}\sigma_m^2).$$

Similarly, the second term tends to

$$\mathcal{N}(0, m^{-1}\mathbb{E}\zeta(1)^2).$$

It is easy to see that the third term tends to zero in probability.

If we now let $m \to \infty$, then

$$\mathcal{N}(0, m^{-1}\sigma_m^2) \to \mathcal{N}(0, \bar{\sigma}^2) \quad \text{in distribution,}$$

where

$$\bar{\sigma}^2 = \lim_{m\to\infty} m^{-1}\sigma_m^2 = \mathbb{E}\zeta(1)^2 + 2\mathbb{E}\zeta(1)\zeta(2),$$

and

$$\mathcal{N}(0, m^{-1}\mathbb{E}\zeta(1)^2) \to 0 \quad \text{in probability.}$$

Consequently,

$$I^{-1/2}\sum_{i=1}^{I}\zeta(i) \to \mathcal{N}(0, \bar{\sigma}^2) \quad \text{in distribution as } I \to \infty.$$

In order to prove (7.15), note first that by the strong law of large numbers for i.i.d. random variables

$$\lim_{N\to\infty}\frac{i(N)m_0}{N} = \lim_{i\to\infty}\frac{i}{T_\alpha(i)} = \pi(s) \quad \text{almost surely.}$$

Hence, for any fixed $\varepsilon > 0$, there exist an integer $N(\varepsilon)$ and a set $\Lambda(\varepsilon) \in \mathcal{F}$ such that $\mathbb{P}(\Lambda(\varepsilon)^c) < \varepsilon$, and for all $N \geq N(\varepsilon)$, $\omega \in \Lambda(\varepsilon)$:

$$\underline{N} \stackrel{\text{def}}{=} [m_0^{-1}(1-\varepsilon)\pi(s)N] + 2 \leq i(N) \leq [m_0^{-1}(1+\varepsilon)\pi(s)N] \stackrel{\text{def}}{=} \bar{N}.$$

Clearly, for such N and ω,

$$\left|\sum_{i=1}^{i(N)}\zeta(i) - \sum_{i=1}^{[m_0^{-1}\pi(s)N]}\zeta(i)\right| \leq 2\max_{\underline{N}\leq j\leq\bar{N}}\left|\sum_{i=\underline{N}}^{j}\zeta(i)\right|.$$

Let $c \geq 0$ be arbitrary. Since any $\zeta(i)$, $i \in \mathcal{I}$, are i.i.d. whenever $\mathcal{I}$ does not contain two consecutive indices, we obtain, by using Kolmogorov's inequality,

$$\mathbb{P}\left\{\max_{\underline{N}\le j\le \bar{N}}\left|\sum_{i=\underline{N}}^{j}\zeta(i)\right|\ge N^{1/2}c\right\}$$

$$\le \mathbb{P}\left\{\max_{\underline{N}\le j\le \bar{N}}\left|\sum_{\substack{\underline{N}\le i\le j\\ i\,\text{even}}}\zeta(i)\right|\ge \tfrac{1}{2}N^{1/2}c\right\}$$

$$+\mathbb{P}\left\{\max_{\underline{N}\le j\le \bar{N}}\left|\sum_{\substack{\underline{N}\le i\le j\\ i\,\text{odd}}}\zeta(i)\right|\ge \tfrac{1}{2}N^{1/2}c\right\}$$

$$\le 4N^{-1}c^{-2}(\bar{N}-\underline{N}+1)\mathbb{E}\zeta(1)^2$$

$$\le 8c^{-2}m_0^{-1}\varepsilon\pi(s)\mathbb{E}\zeta(1)^2.$$

Consequently,

$$\left|N^{-1/2}\sum_{i=1}^{i(N)}\zeta(i)-N^{-1/2}\sum_{i=1}^{[m_0^{-1}\pi(s)N]}\zeta(i)\right|\to 0 \quad \text{in probability as } N\to\infty.$$

It follows that

$$\lim_{N\to\infty}N^{-1/2}\sum_{n=0}^{N}f(X_n)=\lim_{N\to\infty}N^{-1/2}\sum_{i=1}^{i(N)}\zeta(i)$$

$$=\lim_{N\to\infty}N^{-1/2}\sum_{i=1}^{[m_0^{-1}\pi(s)N]}\zeta(i)$$

$$=(m_0^{-1}\pi(s))^{1/2}\lim_{I\to\infty}I^{-1/2}\sum_{i=1}^{I}\zeta(i)$$

$$=\mathcal{N}(0,\sigma^2),$$

where

$$\sigma^2=m_0^{-1}\pi(s)\bar{\sigma}^2$$

$$=m_0^{-1}\pi(s)[\mathbb{E}\zeta(1)^2+\mathbb{E}\zeta(1)\zeta(2)].$$

It remains to calculate the variance $\mathbb{E}\zeta(1)^2$ and the covariance $\mathbb{E}\zeta(1)\zeta(2)$. (To see that the use of Fubini's theorem is justified below the reader is recommended to make the calculations also with $|f|$ instead of f and to use Proposition 5.13.) For the variance we have by (7.13)

$$\mathbb{E}\zeta(1)^2=\mathbb{E}_\nu\left(\sum_{n=0}^{T_\alpha}Z_n\right)^2$$

$$=\sum_{n=0}^{\infty}\mathbb{E}_\nu[Z_n^2;T_\alpha\ge n]+2\sum_{n=0}^{\infty}\sum_{m=n+1}^{\infty}\mathbb{E}_\nu[Z_nZ_m;T_\alpha\ge m]$$

$$=\sum_{n=0}^{\infty}\mathbb{E}_\nu[Z_n^2;T_\alpha\ge n]+2\sum_{n=0}^{\infty}\mathbb{E}_\nu\left[Z_n\sum_{m=n+1}^{\infty}Z_m;T_\alpha\ge n;\right.$$

$$\left.Y_n=\cdots=Y_{m-1}=0\right]$$

$$= \mathbb{E}_\nu \sum_{n=0}^{T_\alpha} \mathbb{E}_{X_{nm_0}}[Z_0^2] + 2\mathbb{E}_\nu \sum_{n=0}^{T_\alpha} \mathbb{E}_{X_{nm_0}}\left[Z_0 \sum_{m=1}^{S_\alpha} Z_m ; Y_0 = 0 \right]$$

by conditioning w.r.t. $\mathscr{F}^X_{nm_0} \vee \mathscr{F}^Y_{n-1}$,

$$= \pi(s)^{-1}\mathbb{E}_\pi[Z_0^2] + 2\pi(s)^{-1}\mathbb{E}_\pi\left[Z_0 \sum_{m=1}^{S_\alpha} Z_m ; Y_0 = 0 \right].$$

Similarly, we get for the covariance

$$\mathbb{E}[\zeta(1)\zeta(2)] = \mathbb{E}_\nu\left[\sum_{m=0}^{T_\alpha} Z_m \sum_{n=T_\alpha+1}^{T_\alpha(1)} Z_n \right]$$

$$= \sum_{m=0}^{\infty} \mathbb{E}_\nu\left[Z_m \sum_{n=T_\alpha+1}^{T_\alpha(1)} Z_n ; T_\alpha \geq m \right]$$

$$= \pi(s)^{-1}\mathbb{E}_\pi\left[Z_0 \sum_{n=1}^{S_\alpha} Z_n ; Y_0 = 1 \right]$$

$$+ \pi(s)^{-1}\mathbb{E}_\pi[Z_0 \zeta(1); Y_0 = 0].$$

By conditioning w.r.t. $\mathscr{F}^X_{m_0} \vee \mathscr{F}^Y_0$ we see that the second term is equal to

$$\pi(s)^{-1}\mathbb{E}_\pi[Z_0 \mathbb{E}_{X_{m_0}}\zeta(1); Y_0 = 0] = 0 \quad \text{by (7.14).}$$

Consequently

$$\sigma^2 = m_0^{-1}\left\{ \mathbb{E}_\pi[Z_0^2] + 2\mathbb{E}_\pi\left[Z_0 \sum_{n=1}^{S_\alpha} Z_n \right] \right\},$$

where

$$\mathbb{E}_\pi Z_0^2 = \mathbb{E}_\pi\left(\sum_{n=0}^{m_0-1} f(X_n) \right)^2 = m_0 \pi(f^2) + 2\sum_{n=1}^{m_0} (m_0 - n)\pi I_f P^n f$$

and

$$\mathbb{E}_\pi\left[Z_0 \sum_{n=1}^{S_\alpha} Z_n \right] = \mathbb{E}_\pi\left[Z_0 \mathbb{E}_{X_{m_0}} \sum_{n=0}^{T_\alpha} Z_n \right]$$

$$= \mathbb{E}_\pi\left[\sum_{n=0}^{m_0-1} f(X_n) G_{m_0,s,\nu} \sum_{m=0}^{m_0-1} P^m f(X_{m_0}) \right]$$

$$= \sum_{n=1}^{m_0} \pi I_f P^n \bar{G}_{m_0,s,\nu} f. \qquad \square$$

Note that, when $m_0 = 1$, the variance $\sigma^2 = \sigma_f^2$ is simply given by

$$\sigma_f^2 = \pi(f^2) + 2\pi I_f P G_{s,\nu} f.$$

If (X_n) has a proper atom α, then

$$\sigma_f^2 = \pi(f^2) + 2\pi I_f U_\alpha f = \pi(f^2) + 2\mathbb{E}_{\pi I_f} \sum_1^{S_\alpha} f(X_n).$$

Corollary 7.3. (i) Suppose that (X_n) is ergodic. The result of Theorem 7.6 holds true for any special function $f \in b\mathscr{L}_0^1(\pi)$.

(ii) Suppose that (X_n) is ergodic of degree 2. Then for any initial distribution λ, any bounded function $f \in b\mathscr{E}$,

$$N^{-1/2}[\xi_N(f) - (N+1)\pi(f)] \to \mathscr{N}(0, \sigma^2_{f-\pi(f)})$$

in $\mathbb{P}_\lambda$-distribution as $N \to \infty$. □

Example 7.4. (*a*) Suppose that (X_n) is an ergodic discrete Markov chain. Then for any two distinct states $x, x_0 \in E_\pi$, any initial distribution λ,

$$N^{-1/2} \sum_{n=0}^{N} (\pi(x)^{-1} 1_{\{X_n = x\}} - \pi(x_0)^{-1} 1_{\{X_n = x_0\}}) \to \mathscr{N}(0, \sigma^2_{x,x_0}),$$

where

$$\begin{aligned} \sigma^2_{x,x_0} &\stackrel{\text{def}}{=} \sigma^2_f \quad \text{with } f = \pi(x)^{-1} 1_x - \pi(x_0)^{-1} 1_{x_0} \\ &= \pi(x)^{-1} + 2\pi(x)^{-1} U_{x_0}(x, x) - \pi(x_0)^{-1}. \end{aligned}$$

If (X_n) is ergodic of degree 2, then for all $x \in E_\pi$,

$$N^{-1/2}\left[\sum_{n=0}^{N} 1_{\{X_n = x\}} - (N+1)\pi(x)\right] \to \mathscr{N}(0, \sigma^2_{1_x - \pi(x)1}),$$

where

$$\begin{aligned} \sigma^2_{1_x - \pi(x)1} &= 2\pi(x)^2 \mathbb{E}_\pi S_x - \pi(x) - \pi(x)^2 \\ &= \pi(x)^3 \mathbb{E}_x S_x^2 - \pi(x). \end{aligned}$$

Notes and comments

Chapter 1

For thorough accounts of the foundations of general (probabilistic) Markov chain theory the reader is referred to Doob (1953), Ch. 5, Neveu (1965), Ch. 5, Orey (1971), Ch. 1, or Revuz (1975), Ch. 0 and 1. Foguel (1969*a*) approaches Markov chains via the theory of contractions on the Banach space $\mathscr{L}^1(\mu)$. Basic references to the theory of discrete Markov chains are e.g. Feller (1957), Chung (1960), Kemeny, Snell & Knapp (1966) and Freedman (1971). The general theory of non-negative operators (from the functional theoretic point of view) is presented in Schaefer (1974) (see also Krasnoselskii, 1964). Monographs concerning non-negative matrices are e.g. Gantmacher (1959) and Seneta (1981). Most of the results treated in our Example (*a*) concerning non-negative matrices can be found in Seneta's book.

Most of the examples treated in this book point out the linkages to other fields having connections with Markov chain theory. On each of these fields (random walks and renewal processes, queueing and storage theory, time series, learning models, branching processes, etc.) there is an extensive literature of its own.

Most of the material of Chapter 1 is of standard character. We make only a couple of detailed comments.

The basic assumption according to which the σ-algebra $\mathscr{E}$ is countably generated is in fact not very restrictive. By using the technique of admissible σ-algebras (see e.g. Orey (1971), Sect. 1.1) one can extend most results to the case where $\mathscr{E}$ is not countably generated.

In order to obtain the minorization inequalities of Section 2.3 we demand for the σ-finiteness of a kernel K that the function f making $\int K(x, \mathrm{d}y) f(x, y)$ finite is jointly measurable (cf. Revuz (1975), Ch. 1, Def. 1.1).

Chapter 2

Many of the basic concepts and results of Chapter 2 go back to Doeblin (1937) and (1940). They were further developed by Doob (1953), Harris (1955, 1956), Orey (1959), Chung (1964), Moy (1967), Jain & Jamison (1967), Isaac (1968), Neveu (1972*b*), Tweedie (1974*a*,*b*) and Revuz (1979). The notions of Chapter 2 are usually formulated for substochastic kernels; however, no additional difficulties arise in the extension to general, non-markovian kernels (see Moy, 1967; Tweedie, 1974*a*,*b*).

Sections 2.1 and 2.2 contain easy preliminary results on closed sets and irreducible kernels. The concept of maximal irreducibility measure was introduced by Tweedie (1974*a*). He also proved Proposition 2.4. Since we do not restrict ourselves to stochastic kernels, we have made a distinction between closed and absorbing sets (see Definition 2.1). So, for example, Orey's (1971) 'closed' means the same as our 'absorbing'.

The small sets, as we call them, are essentially the same concept as the C-sets in Orey (1971). (Orey (1971) assumes – in our notation – that the measure ν is supported by the set C; cf. also our Remarks 2.1.)

Small sets are commonly defined as those sets which satisfy condition (ii) in Corollary 2.1 with $B = E$ (cf. e.g. Foguel, 1969*b*; Brunel, 1971; Lin, 1976). Bonsdorff (1980) observed that these two classes (essentially) coincide (cf. Proposition 2.11(ii)). This implies that the state space splits into a countable number of small sets. The small sets play in many cases the same role as do individual points (or finite sets) in a countable state space.

The fundamental Theorems 2.1 and 2.2 stating the existence of small sets and cycles for irreducible kernels were proved by Jain & Jamison (1967), the presentation of which we closely follow. Many of the arguments used in these proofs go back to Doeblin (1937, 1940) (see also Doob, 1953; Harris, 1956; Orey, 1959; and Isaac, 1968). An extensive study of the concept of cyclicity is made in Chung (1964).

Chapter 3

The potential theoretic results of Sections 3.1 and 3.2 are basically due to Deny (1951) and Doob (1959) (see also Choquet & Deny, 1956). For thorough accounts of the potential theory associated with Markov chains the interested reader should look at Kemeny, Snell & Knapp (1966), Ch. 7–11, and Revuz (1975), Ch. 2 and 6–9.

Theorem 3.2 stating the existence of the convergence parameter and of the R-classification for irreducible kernels is the first basic result in our presentation of the general Perron–Frobenius theory. This theorem was proved for countable non-negative matrices by Vere-Jones (1962, 1967, 1968). The extension to the general state space was performed by Tweedie (1974*a,b*).

A good reference on randomized stopping times is Pitman & Speed (1973).

After some preliminary results (basically due to Doeblin, 1940; Chung, 1964; Moy, 1965*a,b*) concerning the notions of transience and dissipativity we prove in Section 3.5 the fundamental Hopf decomposition result, originally due to Hopf (1954). It is usually formulated in the 'abstract' setting where P is an arbitrary contraction, not necessarily induced by a transition probability, on a space $\mathscr{L}^1(\mu)$ (see e.g. Foguel, 1969*a*, Ch. 2; or Revuz, 1975, Ch. 4).

From Hopf's decomposition theorem we derive the fundamental dichotomy results of Theorems 3.6 and 3.7, due to Jain & Jamison (1967). Most of the other results of Section 3.6 can also be found in Jain & Jamison's paper. Theorem 3.6 is in fact a special case of a more general result, concerning the concept of normal chains (see Jain & Jamison, 1967; or Orey, 1971, Sect. 1.8; cf. also Chung, 1964; Šidak, 1966; Jamison & Orey, 1967; Neveu, 1972*a*; Winkler, 1975; Tuominen, 1976). The important notion of φ-recurrence was introduced by Harris (1956).

The recurrence of autoregressive processes (cf. our Examples 3.4(f) and 5.5(f) is also studied in Athreya & Pantula (1983).

Chapter 4

Sections 4.1 and 4.2 are mostly of standard character. They serve as an introduction to the general regeneration scheme presented in Section 4.3. The concept of regeneration goes back to Palm (1943). It was further developed e.g. by Feller (1949), Bartlett & Kendall (1951) and Smith (1955).

The main result of Chapter 4, Theorem 4.2, was proved independently by Athreya & Ney (1978) and Nummelin (1978*a*), both stimulated by Griffeath (1978). The decomposition results of Theorem 4.1 were formulated for transition probabilities in Nummelin (1978*a*); that they easily extend to non-markovian kernels was noted in Nummelin (1979). The use of Lemma 4.1 appears in Horowitz (1979). The first entrance and last exit decompositions given by Proposition 4.4 and Theorem 4.1 belong to the standard techniques for discrete Markov chains. Proposition 4.7 is adopted from Nummelin (1977). A somewhat different approach to the topics of Section 4.3 is presented in Athreya & Ney (1982).

That the regeneration scheme described in Section 4.4 can be formulated in terms of a certain randomized stopping time ($T_{s,\nu}$ in our notation; cf. Theorem 4.3(i)) is noted in Athreya & Ney (1978). The converse result (part (ii) of Theorem 4.3) was suggested by P. Glynn (personal communication, 1982). Other references related to the theme of Section 4.4 are e.g. Athreya, McDonald & Ney (1978), Nummelin (1978*b*), Berbee (1979), Lindvall (1979*b*), Ney (1981).

The construction of the regeneration time for the many server queues (Example 4.2(*j*)) appears in Charlot, Ghidouche & Hamami (1978). For a construction of regeneration times for tandem queues see Nummelin (1981*b*).

Chapter 5

Theorems 5.1 and 5.2 were proved for finite non-negative matrices by Perron (1907) and Frobenius (1908, 1909, 1912). Perron's and Frobenius' results were extended to countable matrices by Vere-Jones (1962, 1967, 1968). The existence and uniqueness of an invariant measure for a recurrent, discrete Markov chain was proved by Derman (1954); the general, φ-recurrent case was solved by Harris (1956). The proof of the existence and uniqueness of an R-invariant function and measure for a general R-recurrent kernel is due to Tweedie (1974*a*) (for earlier, related works see e.g. Jentzsch, 1912; Krein & Rutman, 1948; Birkhoff, 1957; Karlin, 1959; Harris, 1963). The similarity transform can be found in Harris (1963), Ch. 3, App. 3. The minimality aspect of the R-invariant function h_ν (which leads to Harris recurrent $\tilde{K}$; cf. Proposition 5.4) was noted in Nummelin (1977). The expressions for the R-invariant function and measure in terms of the potential kernel $G^{(R)}_{m_0,s,\nu}$ were given in Nummelin (1979). These constructions are based on the corresponding constructions in the discrete case, due to Derman (1954) and Vere-Jones (1967) (see also Athreya & Ney, 1978, 1982; Nummelin & Arjas, 1976; Nummelin, 1978*a*). The results stated in Theorems 5.1 and 5.2 are also related to the general theory of fixed points for operators in a cone (see Krein & Rutman, 1948). For results concerning subinvariant functions and measures see also Feldman (1962, 1965).

Proposition 5.9 was proved by Harris (1956) (cf. also Kac, 1947). Corollary 5.4 is due to Cogburn (1975). The great practical value of the Balayage principle as a criterion for positive recurrence (and for related concepts) was noted and systematically exploited by Tweedie (1975, 1976) (Proposition 5.10 is taken from Tweedie (1976)). In the discrete case criteria of this kind are often referred to as Foster's (1953) criteria. The rest of Section 5.3 is mostly based on Nummelin (1978*a*). See also Harris (1956), Brunel (1971), Neveu (1972*a*), and Brunel & Revuz (1974) for the notion of special sets and functions. (The special sets are the same concept as the *D*-sets introduced by Harris (1956).) See Isaac (1968) and Cogburn (1975) for the notion of regular states and sets. (Cogburn calls regular sets strongly uniform sets.) Lemma 5.1 is a special case of the general conditional Borel–Cantelli–Levý lemma (see e.g. Doob, 1953, Ch. 7, Sect. 4). For generalized Wald-type identities (cf. Lemma 5.2) see e.g. Franken & Lisek (1982).

Recurrence of degree 2 is extensively studied in Cogburn (1975). He proved Propositions 5.15 and 5.16(i). Most of the other results are special cases of the more general results by Nummelin & Tuominen (1983). For an earlier treatment see Orey (1959).

The presentation of Section 5.5 follows closely Nummelin & Tuominen (1982). The results are generalizations of the corresponding results on the discrete state space, due to Kendall (1959), Vere-Jones (1962) and others. The identities of Lemmas 5.4 and 5.5 are special cases of a general identity by Cogburn (1975). Proposition 5.21 is taken from Nummelin & Tuominen (1982). It extends Popov's (1977) criterion. For a study of the geometric recurrence of reflected random walks (cf. Example 5.5(*d*)) see Miller (1966).

Uniformly recurrent Markov chains are often called strongly recurrent (or ergodic), or also Markov chains having a quasicompact transition probability. They were first studied by Yosida & Kakutani (1941) and Doob (1953). For thorough treatments of uniform recurrence see e.g. Revuz (1975), Ch. 6, §3, and its references. Proposition 5.22 was proved by Horowitz (1972). Proposition 5.23 is basically due to Bonsdorff (1980). Proposition 5.24 is the general counterpart of the corresponding discrete result by Isaacson & Tweedie (1978). Example (*k*) is motivated by Harris (1963), Ch. 3, Sect. 10.

The results of Section 5.7 are straightforward consequences of the corresponding results for transition probabilities.

Chapter 6

Theorem 6.1 is in fact the same as Orey's convergence theorem (Corollary 6.7(i)) for the backward Markov chain $(V_n; n \geq 0)$. The coupling used in its proof is due to Ornstein (1969*a*, *b*). Our presentation follows closely Berbee (1979) (see also Meilijson, 1975). Lemma 6.1 needed in the proof is due to Chung & Fuchs (1951). Theorem 6.2 was proved by Erdös, Feller & Pollard (1949). Theorems 6.3 and 6.4 are due to Pitman (1974); see also Kemeny, Snell & Knapp (1966), Ch. 9. Theorem 6.5 is a special case of a general result by Cogburn (1975). Theorem 6.6 was proved by

Kendall (1959). More general rates for renewal processes are studied in Lindvall (1979*a*). The coupling technique is originally due to Doeblin (1938*a*). Surveys on this method can be found in Griffeath (1978), Berbee (1979) and Thorisson (1981).

Theorem 6.7 was proved for positive Harris recurrent transition probabilities by Orey (1959). The null case was proved in Jamison & Orey (1967). Tweedie (1974*a*) proved the general R-recurrent case. Theorem 6.9 (with $K = P$ Harris recurrent) is due to Horowitz (1979). Theorem 6.10 is essentially due to Jain (1966). More on Orey's theorem and related topics can be found e.g. in Blackwell & Freedman (1964), Jain & Jamison (1967), Jamison & Orey (1967), Horowitz (1969, 1979), Ornstein & Sucheston (1970), Foguel (1971, 1976), Derriennic (1976), Pitman (1976), Revuz (1979), Greiner & Nagel (1982).

Theorem 6.11 is essentially due to Cogburn (1975). The present version is taken from Nummelin (1981*a*). The latter paper also contains more general results on the convergence of sums of transition probabilities and converse results, i.e. criteria for the degrees of recurrence in terms of the convergence of the sums. Theorem 6.12 is from Nummelin (1978*a*); it is the general counterpart of the corresponding discrete results (see e.g. Kemeny, Snell & Knapp (1966), Ch. 9, Pitman (1974)). Theorem 6.13 is a special case of a more general result from Nummelin & Tuominen (1983).

The results of Section 6.5 can be found in Nummelin & Tweedie (1978) and Nummelin & Tuominen (1982). The results in the discrete case were proved by Vere-Jones (1962).

Theorem 6.15 is due to Yosida & Kakutani (1941) and Doob (1953).

Other results on quasi-stationarity are given in Seneta & Vere-Jones (1966) and Tweedie (1974*c*).

The renewal theorem on $\mathbb{R}_+$ (cf. Examples 6.1 and 6.2(*e*)) was proved by Doob (1948), Blackwell (1948) and Smith (1958) (see also Feller, 1971, Ch. 11; McDonald, 1975; Lindvall, 1977, 1979*b*; Arjas, Nummelin & Tweedie, 1978; Ney, 1981). The rate of convergence is studied in Stone & Wainger (1967), Ney (1981), Nummelin & Tuominen (1982, 1983).

Chapter 7

That $\|\sum_0^N P^n f\|$ is a bounded sequence whenever f is a special function satisfying $\pi(f) = 0$ (see Theorem 7.2) is due to Ornstein (1969*a*, *b*), Métivier (1969), Duflo (1969) and Brunel (1971) (see also Foguel & Ghoussob, 1979). Theorem 7.1 improves the results by Duflo (1969), Orey (1971) and Lin (1974*b*). Corollary 7.1 was proved in Nummelin (1978*a*). In fact, the results of Section 7.1 are strongly related to the potential theory of recurrent Markov chains (see Revuz, 1975, Ch. 6–9). The results of Section 7.1 for discrete chains can be found in Kemeny, Snell & Knapp (1966), Ch. 9.

Theorem 7.3 for discrete Markov chains was proved by Doeblin (1938*b*). The present version for general chains is taken from Nummelin (1978*a*). For related works see e.g. Harris (1955, 1956), Chacon & Ornstein (1960), Chacon (1962), Jain (1966), Krengel (1966), Isaac (1967, 1968), Levitan (1967, 1970, 1971), Foguel

(1969*a*, *b*), Métivier (1972), Foguel & Lin (1972), Neveu (1973). Of these Chacon & Ornstein's, Chacon's and Foguel's results are more general in that they deal with the 'abstract' case where P is only assumed to be a contraction on $\mathscr{L}^1(\mu)$.

Theorems 7.4 and 7.5 are taken from Nummelin (1979) (for related results see Orey, 1961, 1971; Kingman & Orey, 1964; Levitan, 1967, 1970; Jain, 1969; Foguel, 1969*b*; Lin, 1974*a*, 1976). A thorough survey of the individual ratio limit theorems is made in King (1981). The regeneration technique is also used in Athreya & Ney (1980).

Theorem 7.6 is a special case of a more general result by Kaplan & Silvestrov (1979). The present proof is adopted from Niemi & Nummelin (1982), which in turn follows the pattern of Chung (1960). Related results can be found also in Doeblin (1937, 1938*b*), Doob (1953), Orey (1959), Cogburn (1972), Grigorescu & Oprisan (1976) and Maigret (1978).

List of symbols and notation

In general, the list below contains only those symbols and notation which are not explained in the main text.

$\mathbb{N}=\{0,1,2,\ldots\}$, the non-negative integers
$\bar{\mathbb{N}}=\mathbb{N}\cup\{\infty\}$, the extended non-negative integers
$\mathbb{R}=(-\infty,\infty)$, the real line
$\bar{\mathbb{R}}=\mathbb{R}\cup\{-\infty\}\cup\{\infty\}$, the extended real line
$\mathbb{R}_+=[0,\infty)$, the non-negative real line
$\mathscr{R}$, $\bar{\mathscr{R}}$, $\mathscr{R}_+$, the Borel subsets of $\mathbb{R}$, $\bar{\mathbb{R}}$, $\mathbb{R}_+$, respectively
$a\vee b=\max\{a,b\}$
$a\wedge b=\min\{a,b\}$
$a_+=a\vee 0$
$a_-=(-a)\vee 0$
$(E,\mathscr{E})$, the basic measurable space (to be called the state space)
$\{f\le g\}=\{x\in E: f(x)\le g(x)\}$ (the notation $\{f<\infty\}$, etc., is interpreted similarly)
$A\backslash B=\{x\in A: x\notin B\}$
$B^c=E\backslash B$
$\sum A_i$, the union of the disjoint sets A_i
ϕ, the empty set
Card (A), the number of elements in the set A
1_A, the indicator of the set A: $1_A(x)=1$ if $x\in A$, $=0$ if $x\notin A$
$A\times B$, the Cartesian product of the sets A and B
$E^{\times n}=E\times\cdots\times E$ (n times)
$E^{\times\infty}=E\times E\times\cdots$
$A\simeq B$, the sets A and B are isomorphic
$\mathscr{A}\cap B=\{A\in\mathscr{A}: A\subseteq B\}$ ($\mathscr{A}$ any collection of sets)
$\mathscr{A}\vee\mathscr{B}=\sigma(\mathscr{A},\mathscr{B})$, the smallest σ-algebra containing the classes $\mathscr{A}$ and $\mathscr{B}$
$\mathscr{A}\otimes\mathscr{B}$, the product σ-algebra of the σ-algebras $\mathscr{A}$ and $\mathscr{B}$
$\mathscr{E}^{\otimes n}=\mathscr{E}\otimes\cdots\otimes\mathscr{E}$ (n times)
$\mathscr{E}^{\otimes\infty}=\mathscr{E}\otimes\mathscr{E}\otimes\cdots$
$\mathbb{E}[\xi|\mathscr{A}]$, the conditional expectation of the random variable ξ w.r.t. the sub-σ-algebra $\mathscr{A}$
$\mathbb{E}[\xi|\mathscr{A}|\mathscr{B}]=\mathbb{E}[\mathbb{E}[\xi|\mathscr{A}]|\mathscr{B}]$
$\mathbb{E}[\xi;A]=\mathbb{E}[\xi 1_A]$

Bibliography

Arjas, E., Nummelin, E. & Tweedie, R.L. (1978). Uniform limit theorems for non-singular renewal and Markov renewal processes. *Journal of Applied Probability*, **15**, 112–25.

Athreya, K.B., McDonald, D. & Ney, P. (1978). Limit theorems for semi-Markov processes and renewal theory for Markov chains. *Annals of Probability*, **6**, 788–97.

Athreya, K.B. & Ney, P. (1978). A new approach to the limit theory of recurrent Markov chains. *Transactions of the American Mathematical Society*, **245**, 493–501.

Athreya, K.B. & Ney, P. (1980). Some aspects of ergodic theory and laws of large numbers for Harris-recurrent Markov chains. In *Colloquia Mathematica Societatis János Bolyai*, **32**. *Nonparametric Statistical Inference*. Budapest, Hungary, 1980.

Athreya, K.B. & Ney, P. (1982). A renewal approach to the Perron–Frobenius theory of non-negative kernels on general state spaces. *Mathematische Zeitschrift*, **179**, 507–29.

Athreya, K.B. & Pantula, S.G. (1983). Strong mixing and Harris recurrence for autoregressive processes. Preprint, Iowa State University.

Bartlett, M.S. & Kendall, D.G. (1951). On the use of the characteristic functional in the analysis of some stochastic processes occurring in physics and biology. *Proceedings of the Cambridge Philosophical Society*, **47**, 65–76.

Berbee, H.C.P. (1979). Random walks with stationary increments and renewal theory. Doctoral thesis. De Vrije University, Amsterdam.

Birkhoff, G. (1957). Extensions of Jentzsch's theorem. *Transactions of the American Mathematical Society*, **85**, 219–27.

Blackwell, D. (1948). A renewal theorem. *Duke Mathematical Journal*, **15**, 145–50.

Blackwell, D. & Freedman, D. (1964). The tail σ-field of a Markov chain and a theorem of Orey. *Annals of Mathematical Statistics*, **35**, 1291–95.

Bonsdorff, H. (1980). Characterizations of uniform recurrence for general Markov chains. *Annales Academiae Scientiarum Fennicae*, Series A, I. Mathematica, Dissertationes, **32**.

Brunel, A. (1971). Chaînes abstraites de Markov vérifiant une condition d'Orey. Extension à ce cas d'un théorème ergodique de M. Métivier. *Zeitschrift für Wahrscheinlichkeitstheorie und verwandte Gebiete*, **19**, 323–29.

Brunel, A. & Revuz, D. (1974). Quelques applications probabilistes de la quasi-compacité. *Annales de l'Institut Henri Poincaré*, B, **10**, 301–37.

Chacon, R.V. (1962). Identification of the limit of operator averages. *Journal of Mathematics and Mechanics*, **12**, 961–68.

Chacon, R.V. & Ornstein, D. (1960). A general ergodic theorem. *Illinois Journal of Mathematics*, **4**, 153–60.

Charlot, F., Ghidouche, M. & Hamami, M. (1978). Irréductibilité et récurrence au sens de Harris des 'temps d'attente' des files *GI/G/q*. *Zeitschrift für Wahrscheinlichkeitstheorie und verwandte Gebiete*, **43**, 187–203.

Choquet, G. & Deny, J. (1956). Modèles finis en théorie du potentiel. *Journal d'Analyse Mathématique*, **5**, 77–135.

Chung, K.L. (1960). *Markov Chains with Stationary Transition Probabilities*. Berlin: Springer.

Chung, K.L. (1964). The general theory of Markov processes according to Doeblin. *Zeitschrift für Wahrscheinlichkeitstheorie und verwandte Gebiete*, **2**, 230–54.

Chung, K.L. & Fuchs, W.H. (1951). On the distribution of values of sums of random variables. *Memoirs of the American Mathematical Society*, **6**.

Cogburn, R. (1972). The central limit theory for Markov processes. In *Proceedings of the Sixth Berkeley Symposium on Mathematical Statistics and Probability*, pp. 485–512. Berkeley, California: University of California Press.

Cogburn, R. (1975). A uniform theory for sums of Markov chain transition probabilities. *Annals of Probability*, **3**, 191–214.

Deny, J. (1951). Familles fondamentales, noyaux associés. *Annales de l'Institut Fourier*, **3**, 73–101.

Derman, C. (1954). A solution to a set of fundamental equations in Markov chains. *Proceedings of the American Mathematical Society*, **5**, 332–34.

Derriennic, Y. (1976). Lois 'zéro ou deux' pour les processus de Markov. Applications aux marches aléatoires. *Annales de l'Institut Henri Poincaré*, B, **12**, 111–29.

Doeblin, W. (1937). Sur les propriétés asymptotiques de mouvement régis par certains types de chaînes simples. *Bulletin de la Société Roumaine des Sciences*, **39**, No. 1, 57–115; No. 2, 3–61.

Doeblin, W. (1938*a*). Exposé de la théorie des chaînes simples constantes de Markov à un nombre fini d'états. *Revue Mathematique de l'Union Interbalkanique*, **2**, 77–105.

Doeblin, W. (1938*b*). Sur deux problèmes de M. Kolmogoroff concernant les chaînes dénombrables. *Bulletin de la Société Mathématique de France*, **66**, 210–20.

Doeblin, W. (1940). Eléments d'une théorie générale des chaînes simples constantes de Markoff. *Annales Scientifiques de l'Ecole Normale Supérieure*, Paris, III Ser., **57**, 61–111.

Doob, J.L. (1948). Renewal theory from the point of view of the theory of probability. *Transactions of the American Mathematical Society*, **63**, 422–38.

Doob, J.L. (1953). *Stochastic Processes*. New York: Wiley and Sons.

Doob, J.L. (1959). Discrete potential theory and boundaries. *Journal of Mathematics and Mechanics*, **8**, 433–58.

Duflo, M. (1969). Opérateurs potentiels des chaînes et des processus de Markov irréductibles. *Bulletin de la Société Mathématique de France*, **98**, 127–63.

Erdös, P., Feller, W. & Pollard, H. (1949). A property of power series with positive coefficients. *Bulletin of the American Mathematical Society*, **55**, 201–04.

Feldman, J. (1962). Subinvariant measures for Markoff operators. *Duke Mathematical Journal*, **29**, 71–98.

Feldman, J. (1965). Integral kernels and invariant measures for Markoff transition functions. *Annals of Mathematical Statistics*, **36**, 517–23.

Feller, W. (1949). Fluctuation theory of recurrent events. *Transactions of the American Mathematical Society*, **67**, 98–119.

Feller, W. (1957). *An Introduction to Probability Theory and Its Applications*, vol. 1, 2nd edn. New York: Wiley and Sons.

Feller, W. (1971). *An Introduction to Probability Theory and Its Applications*, vol. 2, 2nd edn. New York: Wiley and Sons.

Foguel, S.R. (1969*a*). *The Ergodic Theory of Markov Processes*. New York: Van Nostrand.

Foguel, S.R. (1969*b*). Ratio limit theorems for Markov processes. *Israel Journal of Mathematics*, **7**, 384–92.

Foguel, S.R. (1971). On the 'zero–two' law. *Israel Journal of Mathematics*, **10**, 275–80.

Foguel, S.R. (1976). More on the 'zero–two' law. *Proceedings of the American Mathematical Society*, **61**, 262–64.

Foguel, S.R. & Ghoussob, N.A. (1979). Ornstein–Métivier–Brunel theorem revisited. *Annales de l'Institut Henri Poincaré*, B, **15**, 293–301.

Foguel, S.R. & Lin, M. (1972). Some ratio limit theorems for Markov operators. *Zeitschrift für Wahrscheinlichkeitstheorie und verwandte Gebiete*, **23**, 55–66.

Foster, F.G. (1953). On the stochastic matrices associated with certain queueing processes. *Annals of Mathematical Statistics*, **24**, 355–60.

Franken, P. & Lisek, B. (1982). On Wald's identity for dependent variables. *Zeitschrift für Wahrscheinlichkeitstheorie und verwandte Gebiete*, **60**, 143–50.

Freedman, D. (1971). *Markov Chains.* San Francisco: Holden Day.

Frobenius, G. (1908). Über Matrizen aus positiven Elementen I. *Sitzungsberichte der Königlich Preussischen Akademie der Wissenschaften zu Berlin*, 471–76.

Frobenius, G. (1909). Über Matrizen aus positiven Elementen II. *Sitzungsberichte der Königlich Preussischen Akademie der Wissenschaften zu Berlin*, 514–18.

Frobenius, G. (1912). Über Matrizen aus nicht negativen Elementen. *Sitzungsberichte der Königlich Preussischen Akademie der Wissenschaften zu Berlin*, 456–77.

Gantmacher, F.R. (1959). *The Theory of Matrices.* New York: Chelsea.

Greiner, G. & Nagel, R. (1982). La loi 'zéro ou deux' et ses conséquences pour le comportement asymptotique des opérateurs positifs. *Journal de Mathématiques Pures et Appliquées*, **61**, 261–73.

Griffeath, D. (1978). Coupling methods for Markov processes. In *Studies in Probability and Ergodic Theory, Advances in Mathematics, Supplementary Studies*, **2**, 1–43.

Grigorescu, S. & Oprisan, G. (1976). Limit theorems for J-X-processes with a general state space. *Zeitschrift für Wahrscheinlichkeitstheorie und verwandte Gebiete*, **35**, 65–73.

Harris, T.E. (1955). Recurrent Markov processes, II (abstract). *Annals of Mathematical Statistics*, **26**, 152–53.

Harris, T.E. (1956). The existence of stationary measures for certain Markov processes. In *Proceedings of the Third Berkeley Symposium on Mathematical Statistics and Probability*, vol. 2, pp. 113–24. Berkeley, California: University of California Press.

Harris, T.E. (1963). *The Theory of Branching Processes.* Berlin: Springer.

Hopf, E. (1954). The general temporally discrete Markoff process. *Journal of Rational Mechanics and Analysis*, **3**, 13–45.

Horowitz, S. (1969). L_∞-limit theorems for Markov processes. *Israel Journal of Mathematics*, **7**, 60–62.

Horowitz, S. (1972). Transition probabilities and contractions of L_∞. *Zeitschrift für Wahrscheinlichkeitstheorie und verwandte Gebiete*, **24**, 263–74.

Horowitz, S. (1979). Pointwise convergence of the iterates of a Harris-recurrent Markov operator. *Israel Journal of Mathematics*, **33**, 177–80.

Isaac, R. (1967). On the ratio limit theorem for Markov processes recurrent in the sense of Harris. *Illinois Journal of Mathematics*, **11**, 608–15.

Isaac, R. (1968). Some topics in the theory of recurrent Markov processes. *Duke Mathematical Journal*, **35**, 641–52.

Isaacson, D. & Tweedie, R.L. (1978). Criteria for strong ergodicity of Markov chains. *Journal of Applied Probability*, **15**, 87–95.

Jain, N. (1966). Some limit theorems for a general Markov process. *Zeitschrift für Wahrscheinlichkeitstheorie und verwandte Gebiete*, **6**, 206–23.

Jain, N. (1969). The strong ratio limit property for some general Markov processes. *Annals of Mathematical Statistics*, **40**, 986–92.

Jain, N. & Jamison, B. (1967). Contributions to Doeblin's theory of Markov processes. *Zeitschrift für Wahrscheinlichkeitstheorie und verwandte Gebiete*, **8**, 19–40.

Jamison, B. & Orey, S. (1967). Markov chains recurrent in the sense of Harris. *Zeitschrift für Wahrscheinlichkeitstheorie und verwandte Gebiete*, **8**, 41–48.

Jentzsch, R. (1912). Über Integralgleichungen mit positivem Kern. *Journal für die Reine und Angewandte Mathematik*, **141**, 235–44.

Kac, M. (1947). On the notion of recurrence in discrete stochastic processes. *Bulletin of the American Mathematical Society*, **53**, 1002–10.

Kaplan, E.I. & Silvestrov, D.S. (1979). The invariance principle type theorems for recurrent semi-Markov processes with a general state space (English translation). *Theory of Probability and its Applications*, **24**, 536–47.

Karlin, S. (1959). Positive operators. *Journal of Mathematics and Mechanics*, **8**, 907–37.

Kemeny, J.G., Snell, J.L. & Knapp, A.W. (1966). *Denumerable Markov Chains*. Princeton: Van Nostrand.

Kendall, D.G. (1959). Unitary dilations of Markov transition operators and the corresponding integral representation of transition probability matrices. In *Probability and Statistics*, ed. U. Grenander, pp. 138–61. Stockholm: Almqvist and Wiksell.

King, J.H. (1981). Strong ratio theorems for Markov and semi-Markov chains. Doctoral Thesis, University of Wisconsin-Madison.

Kingman, J.F.C. & Orey, S. (1964). Ratio limit theorems for Markov chains. *Proceedings of the American Mathematical Society*, **15**, 907–10.

Krasnoselskii, M.A. (1964). *Positive Solutions of Operator Equations*. Groningen: Noordhoff.

Krein, M.G., & Rutman, M.A. (1948). Linear operators leaving invariant a cone in a Banach space. *Uspehi Mat. Nauk* (N.S.), **3**, 3–95. English translation: *The American Mathematical Society, Translations*, **26** (1950).

Krengel, U. (1966). On the global limit behaviour of Markov chains and of general non-singular Markov processes. *Zeitschrift für Wahrscheinlichkeitstheorie und verwandte Gebiete*, **6**, 302–16.

Levitan, M. (1967). Some ratio limit theorems for a general state space Markov process. Doctoral Thesis, University of Minnesota.

Levitan, M.L. (1970). Some ratio limit theorems for a general state space Markov process. *Zeitschrift für Wahrscheinlichkeitstheorie und verwandte Gebiete*, **15**, 29–50.

Levitan, M.L. (1971). A generalized Doeblin ratio limit theorem. *Annals of Mathematical Statistics*, **42**, 904–11.

Lin, M. (1974*a*). Convergence of the iterates of a Markov operator. *Zeitschrift für Wahrscheinlichkeitstheorie und verwandte Gebiete*, **29**, 153–63.

Lin, M. (1974*b*). On quasi-compact Markov operators. *Annals of Probability*, **2**, 464–75.

Lin, M. (1976). Strong ratio limit theorems for mixing Markov operators. *Annales de l'Institut Henri Poincaré*, B, **12**, 181–91.

Lindvall, T. (1977). A probabilistic proof of Blackwell's renewal theorem. *Annals of Probability*, **5**, 482–85.

Lindvall, T. (1979*a*). On coupling of discrete renewal processes. *Zeitschrift für Wahrscheinlichkeitstheorie und verwandte Gebiete*, **48**, 57–70.

Lindvall, T. (1979*b*). On coupling of continuous time renewal processes. Technical report, University of Göteborg.

Maigret, N. (1978). Théorème de limite centrale fonctionnel pour une chaîne de Markov récurrente au sens de Harris et positive. *Annales de l'Institut Henri Poincaré*, B, **14**, 425–40.

McDonald, D. (1975). Renewal theorem and Markov chains. *Annales de l'Institut Henri Poincaré*, B, **11**, 187–97.

Meilijson, I. (1975). A probabilistic approach to renewal theory. Report BW 53/75, Math. Centre, Amsterdam.

Métivier, M. (1969). Existence of an invariant measure and an Ornstein's ergodic theorem. *Annals of Mathematical Statistics*, **40**, 79–96.

Métivier, M. (1972). Théorème limite quotient pour les chaînes de Markov récurrentes au sens de Harris. *Annales de l'Institut Henri Poincaré*, B, **8**, 93–105.

Miller, H.D. (1966). Geometric ergodicity in a class of denumerable Markov chains. *Zeitschrift für Wahrscheinlichkeitstheorie und verwandte Gebiete*, **4**, 354–73.

Moy, S.T.C. (1965*a*). λ-continuous Markov chains I. *Transactions of the American Mathematical Society*, **117**, 68–91.

Moy, S.T.C. (1965*b*). λ-continuous Markov chains II. *Transactions of the American Mathematical Society*, **120**, 83–107.

Moy, S.T.C. (1967). Period of an irreducible positive operator. *Illinois Journal of Mathematics*, **11**, 24–39.

Neveu, J. (1965). *Mathematical Foundations of the Calculus of Probability.* San Francisco: Holden Day.

Neveu, J. (1972*a*). Potentiel markovien récurrent des chaînes de Harris. *Annales de l'Institut Fourier*, **22**, 85–130.

Neveu, J. (1972*b*). Sur l'irréductibilité des chaînes de Markov. *Annales de l'Institut Henri Poincaré*, B, **8**, 249–54.

Neveu, J. (1973). Une généralisation d'un théorème limite-quotient. In *Transactions of the Sixth Prague Conference on Information Theory, Statistical Decision Functions, Random Processes*, pp. 675–82. Czechoslovak Academy of Science, Prague, 1973.

Ney, P. (1981). A refinement of the coupling method in renewal theory. *Stochastic Processes and their Applications*, **11**, 11–26.

Niemi, S. & Nummelin, E. (1982). Central limit theorems for Markov random walks. *Commentationes Physico-Mathematicae*, **54**, Societas Scientiarum Fennica, Helsinki.

Norman, M.F. (1972). *Markov Processes and Learning Models.* New York: Academic Press.

Nummelin, E. (1977). Strong ratio limit theorems for φ-irreducible Markov chains. *Report HTKK-Mat*-A108, Helsinki University of Technology, Espoo.

Nummelin, E. (1978*a*). A splitting technique for Harris recurrent Markov chains. *Zeitschrift für Wahrscheinlichkeitstheorie und verwandte Gebiete*, **43**, 309–18.

Nummelin, E. (1978*b*). Uniform and ratio limit theorems for Markov renewal and semi-regenerative processes on a general state space. *Annales de l'Institut Henri Poincaré*, B, **14**, 119–43.

Nummelin, E. (1979). Strong ratio limit theorems for φ-recurrent Markov chains. *Annals of Probability*, **7**, 639–50.

Nummelin, E. (1981*a*). The convergence of sums of transition probabilities of a positive recurrent Markov chain. *Mathematica Scandinavica*, **48**, 79–95.

Nummelin, E. (1981*b*). Regeneration in tandem queues. *Advances in Applied Probability*, **13**, 221–30.

Nummelin, E. & Arjas, E. (1976). A direct construction of the R-invariant measure for a Markov chain on a general state space. *Annals of Probability*, **4**, 674–79.

Nummelin, E. & Tuominen, P. (1982). Geometric ergodicity of Harris recurrent Markov chains with applications to renewal theory. *Stochastic Processes and their Applications*, **12**, 187–202.

Nummelin, E. & Tuominen, P. (1983). The rate of convergence in Orey's theorem for Harris recurrent Markov chains with applications to renewal theory. *Stochastic Processes and their Applications*, **15**, 295–311.

Nummelin, E. & Tweedie, R.L. (1978). Geometric ergodicity and R-positivity for general Markov chains. *Annals of Probability*, **6**, 404–20.

Orey, S. (1959). Recurrent Markov chains. *Pacific Journal of Mathematics*, **9**, 805–27.

Orey, S. (1961). Strong ratio limit property. *Bulletin of the American Mathematical Society*, **67**, 571–74.

Orey, S. (1971). *Lecture Notes on Limit Theorems for Markov Chain Transition Probabilities.* London: Van Nostrand.

Ornstein, D. (1969*a*). Random walks I. *Transactions of the American Mathematical Society*, **138**, 1–43.

Ornstein, D. (1969*b*). Random walks II. *Transactions of the American Mathematical Society*, **138**, 45–60.

Ornstein, D. & Sucheston, L. (1970). An operator theorem on L^1 convergence to zero with applications to Markov kernels. *Annals of Mathematical Statistics*, **41**, 1631–39.

Palm, C. (1943). Intensitätschwankungen im Fernsprechverkehr, *Ericsson Technics*, No **44**. Stockholm: Telefonaktiebolaget LM Ericsson.

Perron, O. (1907). Zur Theorie der Matrizen. *Mathematische Annalen*, **64**, 248–63.

Pitman, J.W. (1974). Uniform rates of convergence for Markov chain transition probabilities. *Zeitschrift für Wahrscheinlichkeitstheorie und verwandte Gebiete*, **29**, 193–227.

Pitman, J.W. (1976). On coupling of Markov chains. *Zeitschrift für Wahrscheinlichkeitstheorie und verwandte Gebiete*, **35**, 315–22.

Pitman, J. & Speed, T.P. (1973). A note on random times. *Stochastic Processes and their Applications*, **1**, 369–74.

Popov, N. (1977). Conditions for geometric ergodicity of countable Markov chains. *Soviet Mathematics, Doklady*, **18**, 676–79.

Revuz, D. (1975). *Markov Chains*. Amsterdam: North-Holland.

Revuz, D. (1979). Sur la définition des classes cycliques des chaînes de Harris. *Israel Journal of Mathematics*, **33**, 378–83.

Rosenblatt, M. (1971). *Markov Processes. Structure and Asymptotic Behaviour*. Berlin: Springer.

Schaefer, H.H. (1974). *Banach Lattices and Positive Operators*. Berlin: Springer.

Seneta, E. (1981). *Non-negative Matrices and Markov Chains*, 2nd edn. New York: Springer.

Seneta, E. & Vere-Jones, D. (1966). On quasi-stationary distributions in discrete-time Markov chains with a denumerable infinity of states. *Journal of Applied Probability*, **3**, 403–34.

Šidak, Z. (1966). Classification of Markov chains with a general state space. *Bulletin of the American Mathematical Society*, **72**, 149–52.

Smith, W.L. (1955). Regenerative stochastic processes. *Proceedings of the Royal Statistical Society*, London, A, **232**, 6–31.

Smith, W.L. (1958). Renewal theory and its ramifications. *Journal of the Royal Statistical Society*, B, **20**, 243–302.

Stone, C. & Wainger, S. (1967). One-sided error estimates in renewal theory. *Journal d'Analyse Mathématique*, **20**, 325–52.

Thorisson, H. (1981). The coupling of regenerative processes. Doctoral thesis, University of Göteborg.

Tuominen, P. (1976). Notes on 1-recurrent Markov chains. *Zeitschrift für Wahrscheinlichkeitstheorie und verwandte Gebiete*, **36**, 111–18.

Tweedie, R.L. (1974*a*). *R*-theory for Markov chains on a general state space I: solidarity properties and *R*-recurrent chains. *Annals of Probability*, **2**, 840–64.

Tweedie, R.L. (1974*b*). *R*-theory for Markov chains on a general state space II: *r*-subinvariant measures for *r*-transient chains. *Annals of Probability*, **2**, 865–78.

Tweedie, R.L. (1974*c*). Quasi-stationary distributions for Markov chains on a general state space. *Journal of Applied Probability*, **11**, 726–41.

Tweedie, R.L. (1975). Sufficient conditions for ergodicity and recurrence of Markov chains on a general state space. *Stochastic Processes and their Applications*, **3**, 385–403.

Tweedie, R.L. (1976). Criteria for classifying general Markov chains. *Advances in Applied Probability*, **8**, 737–71.

Vere-Jones, D. (1962). Geometric ergodicity in denumerable Markov chains. *Quarterly Journal of Mathematics*, Oxford, 2nd Ser. **13**, 7–28.

Vere-Jones, D. (1967). Ergodic properties of nonnegative matrices I. *Pacific Journal of Mathematics*, **22**, 361–85.

Vere-Jones, D. (1968). Ergodic properties of nonnegative matrices II. *Pacific Journal of Mathematics*, **26**, 601–20.

Winkler, W. (1975). Doeblin's and Harris' theory of Markov processes. *Zeitschrift für Wahrscheinlichkeitstheorie und verwandte Gebiete*, **31**, 79–88.

Yosida, K. and Kakutani, S. (1941). Operator-theoretical treatment of Markoff's process and mean ergodic theorem. *Annals of Mathematics*, **42**, 188–228.

Index